OXFORD BIOLOGY PRIMERS

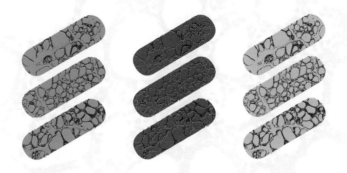

Discover more in the series at

www.oxfordtextbooks.co.uk/obp

Published in partnership with the Royal Society of Biology

CANCER BIOLOGY AND TREATMENT

CANCER BIOLOGY
AND TREATMENT

Aysha Divan and Janice Royds

OXFORD

UNIVERSITY PRESS

Royal Society of
Biology

OXFORD

UNIVERSITY PRESS

Great Clarendon Street, Oxford, OX2 6DP,
United Kingdom

Oxford University Press is a department of the University of Oxford.
It furthers the University's objective of excellence in research, scholarship,
and education by publishing worldwide. Oxford is a registered trade mark of
Oxford University Press in the UK and in certain other countries

Published in the United States of America by Oxford University Press
198 Madison Avenue, New York, NY 10016, United States of America

British Library Cataloguing in Publication Data

Data available

Library of Congress Control Number: 2019954544

ISBN 978-0-19-881347-7

Printed in Great Britain by
Bell & Bain Ltd., Glasgow

Janice Royds would like to dedicate this book to Amy, Zoe, and Lily.

PREFACE

Welcome to the Oxford Biology Primers

There has never been a more exciting time to be a biologist. Not only do we understand more about the biological world than ever before, but we're using that understanding in ever more creative and valuable ways.

Our understanding of the way our genes work is being used to explore new ways to treat disease; our understanding of ecosystems is being used to explore more effective ways to protect the diversity of life on Earth; our understanding of plant science is being used to explore more sustainable ways to feed a growing human population.

The repeated use of the word 'explore' here is no accident. The study of biology is, at heart, an exploration. We have written the Oxford Biology Primers to encourage you to explore biology for yourself—to find out more about what scientists at the cutting edge of the subject are researching, and the biological problems they're trying to solve.

Throughout the series, we use a range of features to help you see topics from different perspectives.

Scientific approach panels help you understand a little more about 'how we know what we know'—that is, the research that has been carried out to reveal our current understanding of the science described in the text, and the methods and approaches scientists have used when carrying out that research.

Case studies explore how a particular concept is relevant to our everyday life, or provide an intimate picture of one aspect of the science described.

The bigger picture panels help you think about some of the issues and challenges associated with the topic under discussion—for example, ethical considerations, or wider impacts on society.

More than anything, however, we hope this series will reveal to you, its readers, that biology is awe-inspiring, both in its variety and its intricacy, and will drive you forward to explore the subject further for yourself.

PREFACE TO CANCER BIOLOGY AND TREATMENT

Cancer is a major health challenge as populations age and the cost of treatment soars. Recent advances in molecular biology have increased our understanding of the complexity of this disease, and efforts are intensifying to translate this into clinical medicine. Current therapies based on the concept of removal of rapidly dividing cells are particularly limited when it comes to treating the spread of disease. Furthermore, cancers can acquire resistance to therapy to which they were previously responsive.

Advances in high throughput technologies that enable interrogation of the genome and associated transcriptional, proteomic, and metabolomic profiles are shedding light on some of these processes. Similarly advances in genetic engineering technologies combined with knowledge of signalling pathways is driving the development of novel strategies such as immunotherapy. It is likely that in the future, patient treatment will become increasingly based on targeting molecular defects that drive a cancer rather on its location.

The field of cancer is dynamic and publications on the subject increase exponentially, so we have had to be selective in the material covered. The primer will take you from a series of core principles, through new findings and hypotheses, and ending with some of the key challenges and opportunities that are being created. In short, we focus on robust principles in conjunction with game-changing findings. A number of references are included at the end of each chapter to broaden your knowledge. The Primer will provide the necessary basis to explore the cancer literature so that you can truly appreciate the dynamism, breadth, and complexity of cancer biology.

ABOUT THE AUTHORS

Dr Aysha Divan, BSc (Hons), PhD

Aysha is an Associate Professor at the Faculty of Biological Sciences University of Leeds. She received her BSc from the University of Leeds and PhD from the University of Sheffield. Aysha's main research interest is in the molecular mechanisms underpinning cancer development and targeting of these processes for therapeutic purposes. She has extensive experience in teaching and curriculum development at undergraduate and postgraduate level, and is the editor and/or author of several texts including *Tools and Techniques in Biomolecular Science* and *Molecular Biology: A Very Short Introduction*.

Dr Janice Royds, BSc (Hons), PhD

Janice is an Honorary Senior Research Fellow in the Department of Pathology at the University of Otago. She received her BSc, PhD, and a Diploma in Higher Education from the University of Sheffield. Janice lectured and carried out research in the Institute for Cancer Studies at Sheffield University for many years before moving to the University of Otago, New Zealand. Her main research interests are the molecular biology of cancer and neurooncology. Janice has a long-standing interest in translational research and enjoys many collaborations with clinical colleagues. Her work has been published extensively in both medical and scientific journals.

ACKNOWLEDGEMENTS

The authors would like to acknowledge Noelyn Hung for her expert advice on pathology, and Eric Blair for his supportive comments and full review of the manuscript.

TABLE OF CONTENTS

1 CANCER FUNDAMENTALS

Learning Objectives

- explain the basis of cancer development as a disease of the genome characterized by the emergence of aberrant cell proliferation and cell survival;
- provide a historical overview of the discovery of the two key classes of cancer genes—the proto-oncogenes and the tumour suppressor genes;
- describe the functions of the common oncogenes and tumour suppressor genes, and how their genetic or epigenetic alteration leads to cancer formation;
- discuss the multi-stage model of cancer development and how our understanding has progressed from a step-wise linear process to a more complex cancer evolution model;
- explain how large-scale cancer genome projects have propelled our understanding of the cancer genome and introduce how this information may be used to manage the disease clinically; and
- describe the role of viruses and inflammation in cancer aetiology.

Cancer is a group of diseases characterized by abnormal growth and survival of cells that have the capacity to invade surrounding tissues and spread (metastasize) to distant organs. An adult human is composed of 10^{15} cells, a number that remains relatively constant. A tight balance between cell proliferation and cell death achieves this. If cell division exceeds cell death, new growth (or 'neoplasia' to give its Greek name) would result. Tumours, from the Latin this time, are swellings comprised of neoplastic cells. Neoplasms can be 'benign' if excessive cell proliferation is localized, or malignant if they invade into surrounding structures. Malignant tumours were given the name 'cancer' due to the crab-like appearance of their invasive projections.

In recent years, our perception of cancer has expanded substantially. With new concepts emerging from research, we now have a better understanding of the molecular changes that drive the morphological and functional cancer **phenotype**. In this chapter, we will introduce the fundamental concepts underpinning cancer development, taking a historical perspective to show how our knowledge of core cancer concepts is evolving.

1.1 Cancer is a disease of the genome

The oldest description of cancer dates back to 1600 BC in Egypt, where eight cases of tumours occurring in the breast and their treatment by cauterization is documented. We know that as early as the eighteenth century, lumps and polyps were surgically removed to reduce deaths from cancer. Later that century, Sir Percivall Pott, an English physician, noticed an association between scrotal cancer in chimney sweeps and exposure to soot. Chimney sweeps in England worked in the nude and did not bathe adequately, resulting in highly carcinogenic hydrocarbons present in soot being retained on the skin of the scrotum. By contrast, chimney sweeps in parts of Europe *did* bathe and *did not* develop scrotal cancer. This led Sir Percivall to recommend that chimney sweep boys wear protective clothing rather than work naked. For the most part, though, cancer and its origins were not comprehended until the advent of molecular biology.

We now know that normal cells evolve to become cancer cells by acquiring successive mutations in primarily two classes of genes: the proto-oncogenes and the tumour suppressor genes. Mutations are alterations to the DNA sequence and can be inherited through the germline as in familial cases of cancer or can occur in somatic cells as in sporadic cases of cancer. Exposure to carcinogens in the environment and lifestyle factors that increase the risk of DNA damage and mutation have been implicated in cancer development. These include chemicals found in tobacco smoke, asbestos, and azo dyes, to name a few. Excessive exposure to UV radiation can also cause cancer, with high frequency UVB rays responsible for 90 per cent of all skin cancers. These and other factors such as diet and their influence on cancer development are discussed in Chapter 3. Some cancers are induced by viruses; the role of viruses in cancer aetiology is described in the section 'Viruses, inflammation, and cancer'.

❯ See more about environmental causes of cancer in Chapter 3.

Genetic defects found in cancers are point mutations, deletions, or insertions of one or a few bases and chromosomal changes (inversion, translocation, and amplification). Epigenetic changes also contribute to cancer formation. These do not directly change the DNA sequence but cause altered gene expression through chromatin remodelling, changes in DNA methylation, and post-translational modification of histone proteins. Both types of changes ultimately lead to the deregulation of a series of biological processes that give rise to distinct phenotypic changes called the 'hallmarks of cancer', a phrase coined by Hanahan and Weinberg in 2000. These include, amongst others, deregulated cell proliferation, resistance to cell death, genomic instability, deregulated metabolism, immune evasion, and activation of invasive and metastatic capabilities, and are discussed further in Chapter 2.

❯ See more about the hallmarks of cancer in Chapter 2.

Proto-oncogenes

Proto-oncogenes code for protein products that are usually involved in signalling pathways promoting cell proliferation, survival, or differentiation. In healthy cells, the level of expression or activity of these proteins is tightly regulated. Genetic or epigenetic changes can convert a proto-oncogene to an oncogene—a term coined by George Todaro and Robert Heuber in 1969. The oncogenic protein product exhibits altered or increased activity driving excessive cell proliferation and inappropriate cell survival thus contributing to tumour formation.

Discovery of oncogenes

Oncogenes were first identified by studying retroviruses. In 1911, Peyton Rous, conducting studies on neoplasia in chickens, showed that retroviruses could induce tumours. Rous prepared cell-free filtrate from chicken sarcoma and injected it into healthy chickens. He found that these chickens also developed tumours and this cycle could be repeated by taking cell filtrate from the new sarcoma and injecting into further healthy chickens. Rous concluded that the causative agent was a virus given that it was small enough to pass through a filter—later coming to be called the Rous Sarcoma Virus (RSV). Decades later, in the 1950s, Harry Rubin showed that the virus could transform normal chicken fibroblasts *in vitro*. Typical characteristics displayed by transformed cells are increased mobility, enhanced proliferation, loss of anchorage dependence, and acquisition of a more rounded morphology. However, it was not until the 1970s that a series of landmark experiments established conclusively that a viral gene caused and maintained the tumour. Experiments using a temperature-sensitive strain of RSV showed that at the lower (permissive) temperature of 37°C the virus was able to transform host cells but not at the higher temperature of 41°C. However, viral progeny production was unaffected. This pointed to the existence of a viral gene responsible for oncogenesis but not required for viral replication. Genetic mapping studies revealed that the viral genome of transformation-competent strains is larger compared to transformation-defective strains. Extra gene sequences were found to be located at the 3′ terminus of the viral RNA genome and designated *V-SRC* (viral sarcoma pronounced v-sark)—thus becoming the first oncogene to be identified. Comparison of DNA transcripts derived from viral infected chicken cells with uninfected chicken cells showed that a gene sequence homologous to the *V-SRC* existed in the genome of un-infected chicken cells. This suggested that the retroviral oncogene was of cellular origin and the retrovirus had at some point incorporated sequences from the chicken host genome into its genome. Subsequent studies in the 1980s showed that *V-SRC* differs from its cellular counterpart *C-SRC* by a C-terminal deletion and by several point mutations. The *SRC* gene encodes a 60 kDa protein with tyrosine kinase activity and is involved in activating signalling pathways leading to cell proliferation. Deletion of the regulatory C-terminal domain and point mutations makes the v-Src protein constitutively active driving inappropriate cell proliferation. By contrast, the protein c-Src has lower kinase activity and negligible oncogenic ability.

The discovery that changes to the *SRC* DNA sequence can convert it to a gene with oncogenic capability suggested that human cancers may also result from mutations in cellular genes. This was confirmed soon after through a study showing that DNA extracted from a human bladder carcinoma cell line could transform the mouse fibroblast cell line NIH3T3 and these transformed cells were able to seed tumours when injected into host mice. The gene responsible for transformation was identified as mutant *H-RAS*, a cellular homologue of *V-RAS* found in Harvey Rat Sarcoma Virus. Since then, a large number of proto-oncogenes have been identified in humans; some of these have viral counterparts whilst others do not.

Mechanisms of activation of proto-oncogenes

As shown in Figure 1.1, there are three main ways by which proto-oncogenes are activated: through point mutations, indels, gene amplification, or chromosomal translocations (or a combination of these). Mutations are dominant in

Figure 1.1 **Mechanisms by which proto-oncogenes are activated.** These include point mutations, indels, gene amplification, and chromosomal rearrangements.

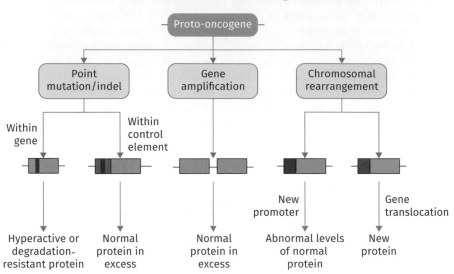

that only one copy of the allele needs to be altered for oncogenic activity to be demonstrated. Table 1.1 lists some common examples of proto-oncogenes.

Point mutations

Point mutations can be missense or nonsense. They can occur in a coding region of a gene resulting in an abnormal protein with enhanced stability or activity. Alternately, point mutations may occur within the regulatory elements causing enhanced or deregulated expression of that gene. An example of a proto-oncogene activated by point mutations is the *RAS* gene. Three *RAS* genes are present in the human genome, *H-RAS*, *K-RAS*, and *N-RAS*. Activating point mutations within these genes predominantly occur in three key codons, codon 12, 13, and 61. For example, a single point mutation (G>T) in the twelfth codon of *RAS* leads to a replacement of the amino acid glycine with valine of the encoded protein product, Ras. Ras proteins are GTPases that cycle between an active GTP-bound form and an inactive GDP-bound form. They transduce upstream signals received from receptor tyrosine kinases (RTKs) to downstream signalling pathways including RAF/MAPK and PI3K/AKT. When activated, these pathways drive cell proliferation. Point mutations reduce the rate of GTP hydrolysis and so mutant Ras remains in an active state driving inappropriate cell proliferation and so contributing to tumourigenesis.

Insertions and deletions (indels)

Another type of DNA alteration are the indels. These refer to the insertion, deletion, or insertion and deletion of nucleotides into the genome. Indels can vary in size ranging from 1 bp to 1 kb in length, and their functional consequence is dependent on whether these changes occur in coding or non-coding regions. Altered gene expression to modified protein function can result. For example, indels in exon 19 of the *EGFR* gene have been identified in non-small cell lung cancers. The EGFR protein encoded by wild-type *EGFR* is a trans-membrane, tyrosine kinase receptor (see Figure 4.1; Chapter 4). It binds epidermal growth

Table 1.1 Examples of oncogenes. For a fuller spectrum of pathways within which the oncogene operates, method of activation, and associated cancers, see COSMIC and/or CBioPortal. RTK = receptor tyrosine kinase

Oncogene	Normal protein function	Example cancer(s)
ABL	Non-receptor tyrosine kinase involved in regulating cell proliferation, apoptosis, differentiation, and migration	Chronic myeloid leukaemia
BCL-2	Anti-apoptotic protein, promotes cell survival	Non-Hodgkin lymphoma
CCND1	Encodes Cyclin D1, a cell cycle regulator that promotes entry into S-phase of the cell cycle to drive cell proliferation	Breast, head, and neck cancers
C-MYC	Transcription factor. Promotes the expression of a wide number of genes involved in multiple cellular processes including differentiation, proliferation, and apoptosis	Ovarian and prostate cancer
EGFR (also known as ERBB1/HER1)	Tyrosine kinase receptor. Activates downstream signalling pathways including RAS/MAPK and PI3K/AKT/mTOR to drive cell proliferation, differentiation, and survival	Glioblastoma / Colorectal cancer
HER2 (also known as ERBB2/NEU)	Tyrosine kinase receptor. Involved in multiple signalling pathways including RAS/MAPK and PI3K/AKT/mTOR and STAT to drive cell proliferation and survival	Breast and ovarian cancers
HDAC1	Histone deacetylase. Removes an acetyl group from the amino acid lysine on histones	Stomach and ovarian cancer
IDH1	Metabolic enzyme isocitrate dehydrogenase 1, involved in cellular metabolism	Lower-grade glioma or secondary glioblastoma
K-RAS	Membrane associated GTPase, transduces signal from RTK to downstream cytoplasmic signal transducers such as Raf	Pancreatic and colorectal cancer
PIK3CA	Catalytic subunit of phosphatidylinositol-3-kinase (PI3K). Acts within the PI3K/AKT/mTOR pathway to promote cell survival, proliferation, and motility	Breast and endometrial cancers
RAF	Cytoplasmic serine-threonine kinase, signal transducer within the RTK/RAS/MAPK signalling pathway	Melanoma
SRC	Non-receptor tyrosine kinase, involved in cell cycle control, proliferation, and survival	Colorectal cancer

factor (EGF) at its extracellular domain, which triggers receptor dimerization, and leads to trans-phosphorylation of its intracellular kinase domains. Activation of downstream signalling and ultimately cell proliferation ensues. The indel in exon 19 alters the ATP binding pocket of the kinase domain of EGFR resulting in appropriate activation of kinase activity and thus increased downstream signalling.

❯ See more about proliferation pathways in Chapter 4.

Gene amplification

In some cancers, the gene is amplified, increasing the number of copies of the gene, which in turn leads to increased expression of the encoded protein. A gene commonly mutated in cancer by this mechanism is *EGFR*. Amplification produces a high concentration of EGFR proteins in the membrane, which leads to high levels of downstream signalling. Another tyrosine kinase receptor, also activated through amplification is *HER2*; this change is observed in a subset of breast cancers.

Chromosomal rearrangements

Some oncogenic changes arise due to chromosomal rearrangements. For example, chronic myeloid leukaemia (CML) is driven by a characteristic chromosomal translocation known as the Philadelphia chromosome (Ph). The *ABL* proto-oncogene is translocated from chromosome 9q to the breakpoint cluster region (BCR) on chromosome 22. This generates a novel tyrosine kinase, Bcr-abl, which is constitutively active, driving cell proliferation. One of the first targeted therapies designed to target oncoproteins was against the Bcr-abl kinase and is described in Chapter 5.

In Burkitt lymphoma, chromosomal translocations between 8 and 24 places the proto-oncogene *C-MYC* under the control of the immunoglobin IgH enhancer leading to constitutive expression of c-Myc. c-Myc is a transcription factor involved in the expression or repression of multiple genes, either preventing or sustaining proliferation. Enhanced c-Myc expression is a major driving force in tumourigenesis, and plays a role in many of the hallmarks of cancer: inducing proliferation, stimulating vascularity, enhancing genomic instability, reprogramming cellular metabolism, and evading immune destruction. Three *MYC* genes are present in the human genome: *C-MYC*, *N-MYC*, and *L-MYC*, and are implicated in the genesis of many types of cancers. Activation of these can be through chromosomal translocation as in the case of Burkitt lymphoma or can be through other mechanisms such as amplification of *N-MYC* observed in neuroblastomas.

Tumour suppressor genes

A second class of genes commonly altered in cancer are the tumour suppressor genes. Tumour suppressor genes code for proteins that play a role in preventing excessive cell proliferation or inhibiting tumour formation, through cellular processes such as cell cycle arrest, apoptosis, and senescence. Tumour suppressor proteins are often expressed in normal cells at very low levels. Their levels rise in response to DNA damage, for example through exposure to radiation, or in response to hyperproliferative signals induced by oncogenes. Unlike proto-oncogenes, both copies of the tumour suppressor gene are usually inactivated to promote tumour development and so tumour suppressors work in a recessive manner. However, some tumour suppressor genes are haploinsufficient—that is, only one of the two copies of the allele needs to be inactivated for the tumour suppressor function to be lost.

The concept of tumour suppression was first put forward by Alfred Knudson in 1971. Knudson theorized that retinoblastomas, a rare childhood cancer of the retina in the eye, develop due to the mutational inactivation of two alleles of the same gene. Based on statistical modelling on the clinical and genetic characteristics of retinoblastoma, Knudson proposed that retinoblastoma can occur in two forms: familial and sporadic. In familial cancer, a child inherits a mutated allele of the retinoblastoma gene, later designated *RB1* from one parent (the first hit). These retinal cells in the infant are heterozygous for the wild-type *RB1* allele. This does not cause a tumour to form, but an inactivating mutation or a deletion of the wild-type second allele of the same gene in retinal cells (second hit) causes loss of heterozygosity and retinoblastoma formation. In non-familial or sporadic cases, two successive mutations inactivate both copies of *RB1* in a single somatic cell. In both instances, tumour development is recessive as both copies of the gene are inactivated for cancer to occur.

Identification of the *RB1* gene

In the 1980s, researchers identified a deletion localized on the long arm (q) of chromosome 13 at band region 14 (13q14) in samples obtained from sporadic and familial cases of retinoblastoma. The DNA sequence corresponding to this locus was cloned by a positional cloning strategy based on linkage analysis of familial retinoblastoma cases and designated *RB1*.

The first evidence that *RB1* is a bona fide tumour suppressor came from a study in mice in which only one allele of the gene was mutated (*Rb+/Rb-*). These mice develop pituitary tumours with nearly 100 per cent penetrance. Analysis of DNA from the cells of these tumours inevitably reveals the loss of the wild-type *Rb* allele, consistent with the two-hit hypothesis and the role of *RB* as a tumour suppressor. Germline mutations of *RB1* predispose humans specifically to retinoblastomas and to a lesser extent to osteosarcomas. Somatic mutations of *RB1* have been found to contribute to malignancy in a wide number of tumours including breast, small-cell lung, bladder, and prostate carcinomas.

The protein encoded by *RB1* is the retinoblastoma protein, pRb. pRb plays a critical role in regulating progression of the cell cycle from the G1 to the S phase. Binding of pRb to the E2F transcription factors restricts the expression of genes that are needed for cell proliferation, leading to cell cycle arrest at the G1/S checkpoint. Functional inactivation of pRb, due to mutation or interaction with viral oncoproteins, relieves pRb repression of E2F and promotes unrestrained entry into S-phase of the cell cycle. See Figure 1.2, which shows how progression through the cell cycle is controlled. The action of pRb is not limited to restricting cell cycle progression but plays a role in multiple cellular processes that are deregulated in cancer; in differentiation, in metabolic pathways and in chromatin organization—emphasizing its importance as a tumour suppressor.

TP53: the second tumour suppressor to be identified

In 1979, a series of researchers working independently described a protein with an apparent molecular mass of 53 kDa, which co-precipitated with the large T antigen of Simian Virus 40 (SV40), a DNA tumour virus. This protein, subsequently named p53, was later found in a variety of human and rodent tumours that had not been infected with SV40. This led to the conclusion that p53 was of cellular rather than of viral origin. In 1984, John Jenkins and colleagues isolated a number of genomic and cDNA clones from tumour cells and showed that primary rat embryo fibroblasts in culture could be transformed in co-operation with the *RAS* oncogene when transfected with the p53 tumour-derived clone. Based on these lines of evidence, by the mid-1980s, p53 was generally considered to be an oncogene.

It took several years to establish that p53 (encoded by *TP53*) was actually a tumour suppressor. In 1989, a new clone of p53 was isolated that could not transform cells. Sequencing of previously used clones showed that earlier studies demonstrating oncogenic properties had utilized mutant forms of p53. So whilst p53 cDNAs carrying mutations could promote cell transformation,

Figure 1.2 Cell cycle progression. Progression through the cell cycle is controlled by the sequential activation of cyclins partnered to specific cyclin dependent kinases (CDK). Cyclin/CDK complexes phosphorylate pRb, which prevents pRb from binding to the transcription factor E2F. E2F is thus free to promote the expression of S-phase genes. When hypophosphorylated, pRB binds to E2F and the cell arrests at the G1/S phase of the cell cycle. Cyclin/CDKs are inhibited by cyclin dependent kinase inhibitors (CDKIs) (p16^{ink4a}, p21^{waf1}), activated in response to e.g. p53 accumulation.

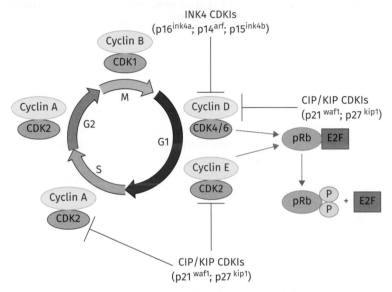

Adapted from Figure 17 in Divan & Royds (2016). Molecular Biology: VSI.

wild-type p53 could not. In fact, overexpression of wild-type p53 was found to suppress *MYC* and *RAS* oncogene induced transformation of cultured cells. Further studies reinforced the role of p53 as a tumour suppressor. Vogelstein and co-workers reported that in human colorectal cancer wild-type p53 activity was lost in a high number of cases through mutations in both alleles of *TP53* and in 1990, *TP53* germline mutations were found in patients with Li-Fraumeni syndrome. This is a rare inherited disorder in which individuals carry one mutated allele of *TP53* through the germline. This predisposes family members to early onset of a wide range of cancers including breast cancer, osteosarcoma, brain tumours, leukaemias, and cancers of soft tissue such as muscle. Whilst the first hit is inherited, the second hit is acquired through somatic mutations resulting in the recessive inactivation of *TP53*. That two hits are required to inactivate tumour suppressor function was further supported by the work of Lawrence Donehower and co-workers in 1992, who showed that mice in which *Trp53* is knocked out develop cancers, mostly lymphomas, with high penetrance.

Approximately half of all human cancers are now known to carry a mutated *TP53* gene and in many more, wild-type p53 activity is deregulated. p53 is a transcription factor that plays a key role in eliminating cells that have either acquired activating oncogenes or excessive genomic damage. The mechanisms by which wild-type p53 guards the genome against inappropriate cell survival are discussed Chapter 4.2.

Mechanisms of tumour suppressor inactivation

Most tumour suppressors are inactivated by deletions or nonsense mutations that result in little or no expression of the respective protein. However, p53 is different: its activity is lost typically through missense point mutations, which usually leads to high levels of inactive protein or a protein with new oncogenic functions. These gain of function mutations are described in Chapter 4. p53, like some other tumour suppressors, can show haploinsufficiency, producing both mutant and wild-type p53 proteins. In such instances, mutant p53 can bind to the wild-type p53 expressed through the intact *TP53* allele and inactivate the wild-type function. Table 1.2 provides a list of common tumour suppressors, their role, and the types of cancers in which they are mutated.

> See more about p53 in Chapter 4.

Mutation patterns between oncogenes and tumour suppressor genes vary, as shown in Figure 1.3. Oncogenes typically have recurrent mutations at localized amino-acid positions. In contrast, mutations in tumour suppressor genes occur scattered throughout the length of the gene, although clustering within specific domains may be evident.

Table 1.2 Examples of tumour suppressor genes. For a fuller spectrum of pathways within which the tumour suppressor acts within, methods of inactivation, and associated cancers, see COSMIC and/or CBioPortal.

Tumour suppressor gene	Normal protein function	Example cancer(s)
APC	Negative regulator of the pro-oncogenic WNT/β-catenin signalling pathway	Colorectal cancer
BRCA1	Protein product interacts with components of the double-stranded DNA repair proteins as part of the DNA damage response.	Breast and ovarian cancers
DPC4 (SMAD4)	Transcription factor involved in the TGF-β signalling pathway	Pancreatic cancer
CDKN2A/INK4A	Encodes p16^{ink4a} a cell cycle dependent kinase inhibitor. Inhibits cell cycle progression	Glioblastoma Melanoma
KMT2D	Encodes a histone methyl transferase, involved in gene regulation through histone modification (H3K4)	Non-Hodgkin lymphoma
PTEN	Encodes a phosphatase, involved in inhibiting proliferation/survival through the AKT/mTOR pathway	Breast, brain, and prostate cancers
PTCH	Encoded protein patched homolog 1 is a transmembrane receptor, an inhibitor of the oncogenic Hedgehog (Hh) signalling pathway	Basal cell carcinoma (skin cancer—non-melanoma)
RB1	pRb regulates cell cycle progression	Retinoblastoma Lung cancer
TP53	p53 transcription factor, induces cell cycle arrest or apoptosis	Lymphomas, breast, lung, and colorectal cancers
VHL	E3 ligase involved in protein degradation e.g. Hif under normal oxygen conditions	Renal cell carcinoma

Figure 1.3 Annotated mutation distribution. (a) Proto-oncogene *K-RAS*. Mutations occur at recurrent positions. (b) Tumour suppressor gene *RB1*. Mutations are scattered throughout the gene.

(a) *KRAS*

(b) *RB1*

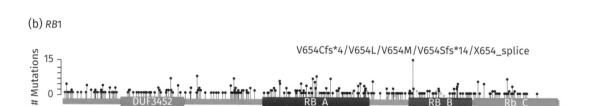

Taken from OncoKB, https://www.oncokb.org.

DNA repair genes

A subset of tumour suppressor genes encodes protein products that act as caretakers of the genome, preventing genomic instability by repairing damaged DNA. If mutations arise in the caretaker genes, this increases the rate at which mutations occur in other genes accelerating the speed at which a normal cell is converted to a neoplastic cell. DNA damage can be due to errors in DNA replication or a result of damage caused by reactive oxygen species (ROS). ROS are produced as by-products of mitochondrial respiration and are able to chemically modify DNA bases through oxidation (e.g. 8-oxoguanine), alkylation (e.g. 6-O-methylguanine), or hydrolysis (deamination, depurination, depyrimidination). In addition to endogenous sources, damage can be through environmental or chemical means. The effects of these vary from formation of pyrimidine dimers and pyrimidine photoproducts (due to UV exposure), to introduction of single or double-stranded DNA breaks (due to ionizing radiation) or cross-linking between bases on the same DNA strand or on adjacent strands (due to chemotherapeutic drugs).

A number of DNA repair systems exist to repair damaged DNA, including nucleotide excision repair (NER), base excision repair (BER), and mismatch repair (MMR), and the pathway(s) activated depend on the type of DNA damage. The basic process for all of these pathways is the excision of damaged regions of the DNA, followed by DNA synthesis to fill the gap and subsequent ligation of the fragments. One of the earliest set of researchers to document a biological repair process in mammalian cells was Ronald Rasmussen and Robert Painter. In 1964, they showed that non-semiconservative DNA synthesis occurred in UV radiation-induced damage to DNA. This was subsequently supported by studies conducted on skin biopsies derived from normal skin fibroblasts and those from patients with the hereditary disease xeroderma pigmentosum (XP). XP is a rare autosomal recessive disease in which patients show a sensitivity to sunlight and are predisposed to skin cancer. This work showed that normal skin fibroblasts can repair UV radiation-induced damage to DNA by inserting new bases into the DNA, but cells from XP patients were unable to repair or showed a diminished repair response. We now know that DNA excision repair proteins such as XPA, XPB, and XPC involved in the NER pathway are defective in XP patients. These individuals have a 1,000-fold increased risk of developing skin cancer and a >10–20-fold increased risk of other cancers such as leukaemia, and brain and lung tumours before the age of twenty. NER is used to excise pyrimidine dimers and other types of bulky DNA damage that distort the helix whilst BER is used for excising uracil and other misincorporated bases or excising bases that have undergone simple damage such as oxidation.

A second example of a defective DNA repair pathway in causing cancer came from studies conducted in the 1990s. Researchers showed that patients with hereditary non-polyposis colon cancer (HNPCC), also known as Lynch syndrome, was caused by mutations in the *MSH2* and *MLH1* mismatch repair genes. MMR operates to correct mistakes generated during DNA replication and thus ensures DNA is replicated faithfully. Errors in DNA replication are corrected through DNA polymerase proofreading processes or if missed by proofreading, corrected by MMR. Mutations in one of four MMR genes—*MLH1*, *MSH2*, *MSH6*, or *PMS2*—increases an individual's risk of developing HNPCC and also other tumours such as endometrial, ovarian, gastric, and small intestinal cancers. In HNPCC, defects in MMR lead to microsatellite instability (MSI). Microsatellites are short DNA sequences repeated many times in tandem and can be located in coding and non-coding regions of the genome. These regions are prone to DNA polymerase slippage during DNA replication and if not corrected by MMR leads to lengthening or shortening of the sequence. If located in a protein-coding region, frame-shift mutations can arise resulting in altered or non-functional proteins. MSI generates a mutator phenotype, increasing substantially the rate at which mutations accumulate in the genome and so cancers with MSI carry a very high mutational burden. Cancers with defective proofreading exonuclease domains of the replicative DNA polymerases POLD and POLE also exhibit a mutator phenotype and correspondingly an excessive mutational burden. Genes encoding these tumour suppressors are mutated in a subset of lung, endometrial, and colorectal cancers.

Damage to DNA can also result in single or double-stranded DNA breaks. Double-stranded breaks (DSBs) are the most threatening form of DNA damage and if unrepaired can lead to chromosomal rearrangements or cell death. DSBs are repaired by either homologous recombination (HR) or non-homologous end joining (NHEJ). Mutations in *BRCA1* and *BRCA2* are involved in the HR pathway and germ-line mutations in these genes are associated with a subset of familial breast and ovarian cancers. These and other cancers arising from genetic defects in DNA repair systems are discussed further in Chapters 3 and 4.

❯ See more about DNA repair mechanisms in Chapters 3 and 4.

Epigenetic alterations in cancer

Whilst both oncogenes and tumour suppressor genes are altered by genetic changes, their expression can also be increased or decreased by epigenetic changes. A well-characterized epigenetic modification is DNA methylation in which a methyl group is added to a cytosine nucleotide adjacent to a guanine (CpG where p represents a phosphate) of DNA. This change usually occurs within regions of the genome with a high density of CpG dinucleotides, termed CpG islands, and has been documented for various tumour suppressor genes including *CDKN2A* and *BRCA1*. Methylation in these cases leads to gene silencing and therefore loss of tumour suppressor activity. Oncogenes can be inappropriately activated through removal of methyl groups. For example, promoter hypomethylation, as observed in *RAS*, leads to gene expression and increased oncogenic activity.

Other epigenetic changes that contribute to changes in gene expression include posttranslational modification of histone proteins through acetylation, methylation, phosphorylation, or ubiquitylation. Modification of histones, leads to a more open or closed chromatin structure, thereby increasing or decreasing accessibility of transcription factors to the target gene promoter. These in turn activate or repress gene expression. An example is the phosphorylation of histone 3 at amino acid serine 10 (H3S10ph) and of histone 2B at serine 32 (H2BS32ph). These are markers of transcriptional activation and are known to be involved in the inappropriate expression of the proto-oncogenes *MYC*, *JUN*, and *FOS*. In contrast, trimethylation of histone 3 at lysine 27 (H3K27me3), a marker of transcriptional repression is involved in silencing the expression of some tumour suppressor genes.

The addition or removal of chemical groups is catalysed by different enzymes including methyl transferases, histone acetyl transferases, and histone deacetylases. Deregulation of these enzymes leads to altered epigenetic states and inappropriate oncogene or tumour suppressor gene expression. Epigenetics and its role in cancer are discussed further in Chapters 4 and 5 (see also the section 'Identification of novel cancer-causing genes' and Tables 1.1 and 1.2).

❯ See more about epigenetics in Chapters 4 and 5.

 Key Points

- Cancers arise as a consequence of genetic and epigenetic changes in two key classes of genes: the tumour suppressor genes and the proto-oncogenes.
- Both types of changes ultimately lead to the deregulation of a series of biological processes that give rise to distinct phenotypic changes termed the hallmarks of cancer.

1.2 Multi-stage model of cancer development

Normal cells evolve progressively to become cancer cells over time. This process, termed tumour progression, requires multiple genetic and epigenetic changes in critical genes that alters their expression and confers an enhanced growth and survival advantage to that cell and to its descendants. These changes take time to accumulate and so cancer is primarily a disease of ageing. This is supported by epidemiological studies that show that the incidence of cancer increases with age (see Chapter 3). Cancer development as a multistage process was first proposed by Isaac Berenblum and Philip Shubik in the 1940s based on studies showing that mouse skin cells exposed to mutagenic carcinogens remained latent until further outgrowth is promoted by subsequent treatment with non-mutagenic promoters such as phorbol esters. This led to the formulation of the initiator-promoter model of cancer development. In this model, cancers are thought to arise from a single founder cell that acquires a mutation randomly. This initiating mutation confers a growth or survival advantage to the cell and its descendants, and so the cell proliferates more effectively than its neighbours. One of the mutant cells in this expanded clone then acquires a second promoting mutation causing a second cycle of clonal expansion. This process of clonal expansion then repeats itself, each triggered by a specific genetic or epigenetic change ultimately leading to a large tumour mass containing billions of cells.

That tumours arise from a single mutated cell, accumulating additional mutations as it progresses was advanced by Vogelstein and colleagues in the 1980s. Characterizing the development of colorectal cancers (CRC), these authors proposed that the first initiating mutation most often arises in the *APC* gene, a tumour suppressor that confers a selective growth advantage resulting in a small **adenoma**. The adenoma grows slowly but a second promoting mutation in another gene such as *K-RAS* leads to a second round of clonal growth that allows an expansion of cell number. This process of mutation followed by clonal expansion continues with mutations in further genes such as *TP53* and *PI3K*. With an increasing number of mutations, the tumour evolves from benign to malignant progressing from small adenoma, to mid adenoma, to large adenoma, and eventually to **carcinoma**. This linear, step-wise **clonal evolution** model does not explain the complex clonal heterogeneity observed in tumours and alternative tumour evolution models are now being identified. We will return to these later, when tumour intra-heterogeneity is discussed.

❯ See more about the influence of age on cancer incidence in Chapter 3.

1.3 Cancer genomics

Large-scale sequencing projects including The Cancer Genome Atlas (TGCA) launched in the USA in 2006 and the Cancer Genome Project launched in the UK in 2010 have caused a substantial shift in the way we understand cancer. These projects have characterized the genomes of tumours from almost all cancer types using **exome** or **whole genome sequencing**. The huge volume of data generated is stored in the COSMIC database. Box 1.1 lists this and other databases that catalogue this huge volume of data. Whilst challenging to interpret, data from these projects have provided deep insights into the nature of the cancer genome and in turn, are informing how we treat and manage the disease. Figure 1.4 provides an overview of how this information is being applied in a research and clinical setting.

Sources of large-scale cancer genome data

- The Cancer Genome Atlas (TCGA). Maintains a database of key genomic changes in thirty-three cancer types. https://cancergenome.nih.gov
- The International Cancer Genome Consortium. Has the goal of assembling a comprehensive catalogue of genomic, transcriptomic, and epigenomic alterations in fifty cancer types. https://icgc.org
- Cancer Genome Project. Maintains COSMIC, a Catalogue of Somatic Mutations in Cancer curated from the published literature and international consortia. https://cancer.sanger.ac.uk/cosmic
- OncoKB. Oncology Knowledge Base contains information about the effects and treatment implications of specific cancer gene alterations, curated from published literature and other sources. https://www.oncokb.org
- cBioPortal for cancer genomics. Allows visualization, analysis, and download of large-scale cancer genomics data sets. http://www.cbioportal.org/index.do
- GeneCards. Integrates human gene data from ~125 web sources. https://www.genecards.org

Figure 1.4 Key findings from large-scale cancer genome projects and potential implications for basic research and clinical management.

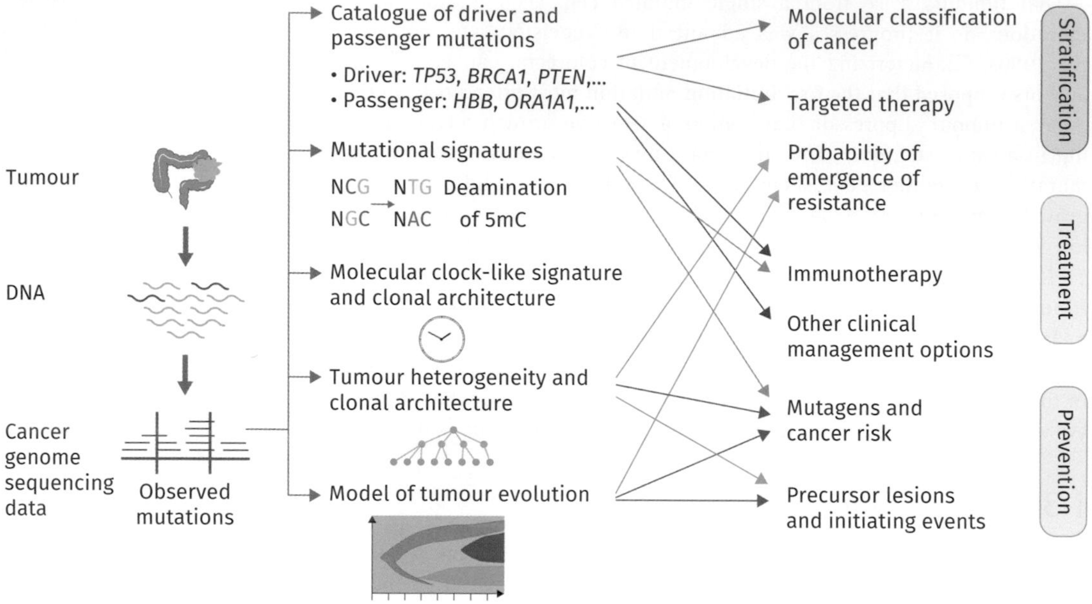

Adapted from Figure 1 in De & Ganesan (2017). Annals of Oncology, 28, 938–45.

Identification of novel cancer-causing genes

In sequencing projects, genomes of samples derived from tumour cells and those derived from healthy tissue from the same patient are compared. The purpose is to identify genetic abnormalities such as DNA mutations, chromosomal translocation, and copy number variations (gene amplifications, deletions, or duplications), as well as changes in gene expression and in epigenetic modifications. Such studies have led to the identification of new cancer-causing genes, some encoding proteins involved in well-established signalling pathways linked to proliferation and survival but others less well-studied in cancer. An example is the discovery of mutations in *IDH1*. *IDH1* encodes the cytoplasmic enzyme isocitrate dehydrogenase responsible for converting isocitrate to α-ketoglutarate in the TCA cycle. Mutations in *IDH1* and its mitochondrial homologue *IDH2* confer a gain of function activity whereby isocitrate is converted to the oncometabolite 2-hydroxyglutarate (2-HG). 2-HG promotes tumour development by inhibiting a number of enzymes that require α-ketoglutarate as a substrate. *IDH1* and *IDH2* mutations identified in gliomas and in acute myeloid leukaemia (AML) were the first genes linking cellular metabolism to cancer. Deregulated metabolism has become increasingly recognized as an important hallmark of cancer cells.

Another example was the finding that genes encoding epigenetic and chromatin regulators are frequently mutated in cancer. One such example is genes encoding components of the nucleosome-remodelling complex SW1/SNF. One of these, *PBRM1*, which encodes BAF180, is frequently inactivated in renal cancer. Wild-type BAF180 is able to induce the expression of genes involved in cell cycle arrest and apoptosis such as *CDKN1A* and *TP53* by increasing accessibility of transcription factors to their promoters. Loss of BAF180 silences the expression of these critical promoters, leading to aberrant cell proliferation.

❯ See more about deregulated metabolism and the role of *IDH* in Chapter 4.

Cancer driver mutations

When the genomes of cancers of the same type from different individuals are compared, some genes appear to be frequently mutated across the tumours whilst others are mutated less frequently. A small number of key susceptibility genes such as *TP53*, *APC*, and *BRAF* display mutations at high frequency. These are termed driver mutations and confer a selective growth advantage, leading to the outgrowth of a tumour clone. The rest, called passenger mutations are more randomly distributed across the genome and their involvement in promoting cancer progression is less clear. For example, exome-analysis of 100 breast cancer genomes in 2012 identified 250 driver mutations (somatic mutations and/or copy number variations) located within forty cancer genes. Fifty-eight per cent of the driver mutations were concentrated in seven key genes (*TP53*, *PIK3CA*, *ERBB2 (HER2)*, *MYC*, *FGFR1*, *GATA3*, and *CCND1*). Subsequent studies examining the whole genome (protein and non-protein coding regions) have identified further genes that carry driver mutations associated with breast cancer, for example *FOXP1*. Driver mutations located in non-protein coding regions of the genome have also been identified such as splice sites mutated in *GATA3* and *TP53*.

The number of mutated genes driving cancer varies between tumour-types. A recent analysis identified kidney chromophobe (a rare type of kidney cancers) as having the fewest cancer **driver genes** (two genes: *TP53* and *PTEN*) and endometrial cancers as having the most (fifty-five genes). Typically, between two and ten alterations in driver genes are required to yield a transformed phenotype and these tend to occur in different signalling pathways. This co-operation ensures that the cancer cell gains the greatest proliferative advantage.

Although the number of driver genes in any given cancer can be small, combined with the passengers, a tumour can carry hundreds to thousands of mutations. Some cancers such as acute leukaemias harbour fewer mutations, around 1,000 whilst others such as lung cancer and melanoma can harbour over 100,000 mutations. The reasons for this difference in mutational burden is not clear but exposure to powerful mutagens such as UV light and tobacco carcinogens may explain the higher rates observed in skin and lung cancers as would MSI in cancers with defective mismatch repair mechanisms.

Comparison between tumours: inter-tumour heterogeneity

Studies mapping the mutational landscape of major cancer types highlight that substantial mutational heterogeneity exists between tumours of the same organ-type. This inter-tumour heterogeneity has led to the categorization of tumours into molecular sub-types based on patterns of genetic and epigenetic alterations found in tumours. Alterations are grouped based on the cellular pathway within which it operates. For example, changes in *EGFR*, *RAS*, or *RAF* lead to activation of cell proliferation through a deregulated RTK/RAS/MEK pathway and so these oncogenes are categorized under the RTK-RAS signalling pathway. Similarly, loss of *RB1* leads to a loss of cell cycle arrest and so *RB1* alterations are categorized under the cell cycle pathway. Based on this approach, breast cancers for example, are divided into four molecular subtypes; hormone-receptor (HoR) positive for either the oestrogen or progesterone receptor (ER+ and/or PR+) which are further subdivided into luminal A and luminal B based on gene expression patterns, HER2+/HoR−, and triple negative (HER2−/HoR−) breast cancers.

Figure 1.5 Four breast cancer subtypes defined from the molecular analyses of breast cancer patients. (a) Consensus clustering analysis identified four major breast cancer types (samples n = 5348). (b) Heat map defining the four molecular subtypes. The red bar indicates membership of a cluster type. Results integrated from the analysis of five different genomic/proteomic platforms including miRNA sequencing, DNA methylation, DNA copy number (CN) array, PAM50 mRNA array (a fifty-gene expression panel), and reverse phase protein array (RPPA) (expression of 171 cancer-related proteins and phosphoproteins). (c) Association with molecular (mutation status of significantly mutated genes *TP53, GATA3, PI3KCA, MAP3K1, MAP2K4*) and clinical (T = tumour size and N = node status) features.

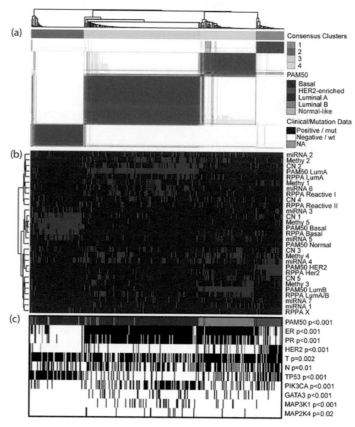

Figure 2 in The Cancer Genome Atlas Network (2012). Nature, 490, 61–70.

Figure 1.5 shows the four breast cancer subtypes defined from the molecular analyses of breast cancer patients. Molecular subtypes correlate with clinical outcomes and can be used to aid prognosis and guide treatment approach as discussed in Chapters 5 and 6.

❯ See Chapter 5 for cancer treatments and Chapter 6 for the role of biomarkers in cancer therapy.

Comparison across cancers

Comparisons of tumours arising from different organs have shown that molecular alterations can be shared across tumour types. A 2013 TCGA study comparing 12 cancer types showed that of the 127 significantly mutated genes (SMGs) studied, some mutations were associated with multiple cancers, whilst others were limited to one cancer type. The significantly mutated genes were categorized under twenty broad cellular processes based on the pathways in which they were involved (see Figure 1.6). Notably, driver genes categorized under the cellular processes genome integrity (*TP53*), PI3K signalling (*PIK3CA* and *PTEN*), RTK signalling (*K-RAS*), and histone modifiers (*ARID1A*) associated with more than one tumour type whilst those grouped under Wnt/β-catenin signalling (APC) appeared less frequently. A more recent study looking at larger sets of driver genes (258) across thirty-three cancer types have shown similar results.

Figure 1.6 Comparison of significantly mutated genes (SMGs) across twelve cancer types. A total of 127 genes identified as significantly mutated were classified under one of twenty cellular processes within which it operates. Percentages of samples mutated in individual tumour types and Pan-Cancer (across the twelve tumours) are shown, with the highest percentage in each gene amongst the twelve cancer types highlighted in bold.

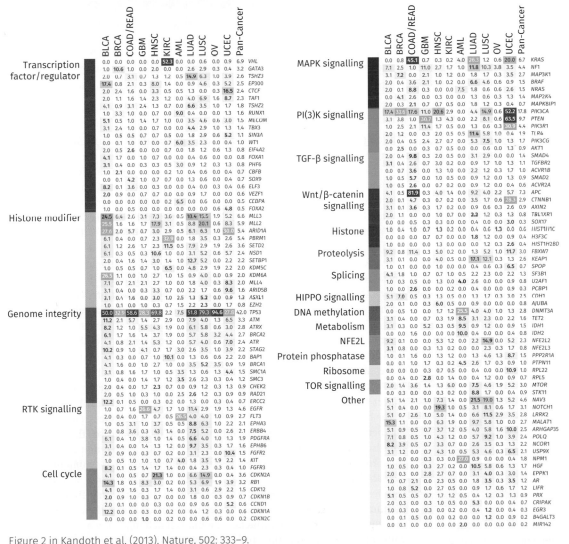

Figure 2 in Kandoth et al. (2013). Nature, 502: 333–9.

One hundred and forty-two out of 258 drivers were associated with a single cancer whilst eighty-seven driver genes had roles in two or more cancer types. *TP53* was shown to be the most commonly shared driver (across twenty-seven cancer types), followed by *PIK3CA*, *KRAS*, *PTEN*, and *ARID1A*, each of which was associated with fifteen or more cancer types. Such findings have led to a revision of how tumours are classified. Traditionally tumours were classified based on the histopathological characteristics of the tumour (see Chapter 2), but this information is now integrated with that derived from molecular analysis. This revised approach has important implications for cancer therapy, as it means that patients can be stratified based on the molecular profile of their tumour and drugs targeting common mutations can be administered across multiple cancer-types. Therapy and patient stratification are topics that we return to in Chapters 5 and 6.

❯ See Chapters 5 and 6 for therapy and patient stratification, respectively.

> ### 💡 Key Points
>
> - Most human cancers are caused by two to ten alterations in driver genes that are acquired over a period of twenty to thirty years.
> - Driver mutations tend to occur in different signalling pathways. This co-operation ensures that the cancer cell gains the greatest proliferative advantage.

Intra-tumour heterogeneity

Although tumours are thought to be clonal in that the tumour cells derive from a single founder cell, all the cells within a single tumour are not identical. This would be expected if tumour progression followed a linear scheme of successive driver mutations arising in a clone which expanded and replaced the less fit clones in the tumour. Instead, tumours are composed of multiple sub-populations with distinct genetic and epigenetic profiles. This diversity within the primary tumour is termed intra-tumour heterogeneity. How does this intra-tumour heterogeneity arise? One mechanism proposed is genomic instability. As tumour cells acquire multiple genetic changes, the genome becomes unstable and so the rate at which mutations are acquired is elevated. This means that new clones are unable to replace their less fit predecessors before further changes accumulate driving their expansion. Distinct cell populations thus arise that share a number of genetic and epigenetic changes, but then diverge in the number and type of mutations that subsequently accumulate. Evidence for this clonal diversification and more complex evolution of cancers comes from genome-sequencing studies. Scientific Approach Panel 1.1 shows how intra-tumour heterogeneity can be reconstructed by genome analyses of cancer samples obtained from multiple regions within individual tumours.

The heterogeneity observed between tumour cells is also shaped by the tumour microenvironment (TME). In a solid tumour, the TME comprises the cancer cells but also other cells, including endothelial cells, fibroblasts, and immune cells supported by a vascular system. Regional differences in the tumour such as availability of oxygen can impose different selective pressures on the tumour cells leading to subclones with different molecular profiles in different areas of the tumour. Paracrine signalling between tumour cells and their neighbouring non-tumour cells can also influence tumour cell heterogeneity as can treatment with cancer drugs. Intra-tumour heterogeneity poses a significant challenge to tumour therapy to which we will return in Chapter 6.

❯ See more about tumour heterogeneity in Chapter 6.

Scientific approach panel 1.1
Reconstruction of intra-tumour heterogeneity in kidney cancer

Clear cell kidney cancer (CCRCC) accounts for 75 per cent of all kidney cancer cases and occurs in both inherited and sporadic forms. In both types of ccRCC, the *VHL* tumour suppressor gene is inactivated in approximately 90 per cent of cases through either genetic mutation or hypermethylation. To identify intra-tumour heterogeneity and thus better understand how the tumour evolves, Gerlinger and colleagues in 2012 conducted multi-region exome sequencing of tissue samples (primary and associated metastases) obtained from two ccRCC cancer patients. This study was extended in 2014 with samples derived from a larger number of cancer patients. For simplicity, data from one of the patients is shown in Figure SA 1.1. In this patient, approximately one-third of the 128 validated mutations were present in all of the

regions sampled and of these only one driver gene *VHL*, was mutated in all of the regions. In a phylogenetic tree constructed from the sequencing data, this was designated a truncal mutation, an early event in the evolution of the cancer. Other driver mutations including *SETD2* (encoding a histone methyltransferase), *mTOR* (encoding the mammalian target of rapamycin kinase), and *KDM5C* (encoding a histone demethylase) were confined to spatially separated subclones and were designated branch mutations. The authors concluded that tumour evolution in ccRCCs follows a branched model in which each subclone carries a distinct spectrum of mutations with the exception of the ubiquitous VHL truncal mutation. Differences also existed between the primary tumour and the metastatic samples. Cells that seeded

Figure SA 1.1 Reconstructing intra-tumour heterogeneity in renal cell carcinoma. (A) Biopsies harvested from regions of a primary renal carcinoma (R1 to R9; G indicates tumour grade) and from metastases M1 (perinephric) and M2A and M2B (chest wall). (B) Heat map showing regional distribution of 101 somatic point mutations and thirty-two indels (light grey shading means the mutation is present and dark shading means mutations is absent) in seven primary tumour regions and three metastases. The colour bars above the heat map indicate classification of mutations according to whether they are ubiquitous, shared by primary-tumour regions, shared by metastatic sites, or unique to the region (private). Among the gene names, purple indicates that the mutation was validated, and orange indicates that the validation of the mutation failed. (C) Phylogenetic relationships of the tumour regions. R4a and R4b are the subclones detected in R4. A question mark indicates that the detected *SETD2* splice-site mutation probably resides in R4a, whereas R4b most likely shares the *SETD2* frameshift mutation also found in other primary-tumour regions. Branch lengths are proportional to the number of somatic mutations separating the branching points. Potential driver mutations were acquired by the indicated genes in the branch (arrows).

(a)   Biopsy site

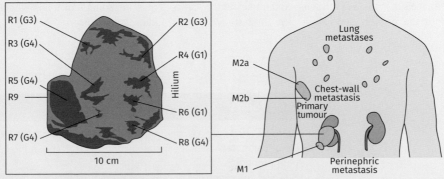

Adapted from Figure 2 in Gerlinger et al. (2012). New England Journal of Medicine, 366: 883–92

Continued

Figure SA 1.1 Continued

(b)   Regional distribution of mutations

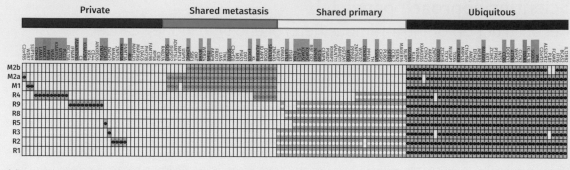

(c)   Phylogenetic relationships of tumour regions

- Ubiquitous
- Shared primary
- Shared metastasis
- Private

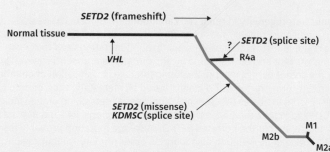
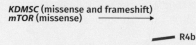

metastases had diverged at an early stage and lacked the *mTOR* mutation. These cells also showed different mutations in *SETD2* (missense instead of frameshift) and in *KDM5C* (splice site instead of missense and frameshift) compared to the primary sub-clones. The 2012 study was the first to provide a detailed understanding of intra-tumour heterogeneity. Since then a branched model of cancer evolution has been identified in a number of cancer types including leukaemia, glioblastoma, breast, colon, and prostate cancers.

Key Points

- Tumours arise from a single transformed cell via the accumulation of multiple driver mutations that generate genotypes that confer increasingly malignant properties.
- However, the genotypes of cells that make up a tumour are not all the same. Instead, the tumours comprise of a collection of clones with different genetic and phenotypic profiles.

1.4 Cancer stem cells

Cancer stem cells have also been put forward as a way of explaining intra-tumour heterogeneity. In any given tissue, cells are organized in a hierarchy starting with stem cells at the top. This stem cell undergoes asymmetrical cell division to form two daughter cells: one remains a stem cell capable of self-renewal and the second is a transit-amplifying cell, also called a progenitor cell. Following many successive rounds of symmetrical cell division, these progenitor cells subsequently undergo differentiation to give rise to the highly specialized cells of that tissue. Differentiated cells are short-lived and frequently shed from the tissue. In contrast, stem cells and progenitor cells are longer-living and therefore able to accumulate the sequential genetic and epigenetic changes required for tumour formation. It is now widely accepted that in some haematological and solid cancers mutations arise either in stem cells forming cancer stem cells or in the progenitor cells. It is possible for a mutated progenitor cell to become dedifferentiated to become a cancer stem cell. Thus, tumour heterogeneity can be a consequence of mutations occurring in progenitor cells at different stages of differentiation or in stem cells. Clonal evolution and the cancer stem cell model are not mutually exclusive; cancer stem cells may also undergo clonal evolution, resulting in stem cells with divergent mutation profiles and increased tumour heterogeneity. All three models are shown in Figure 1.7.

For a cell to be defined as a cancer stem cell it must be capable of self-renewal and be able to drive tumour growth. A method for testing the presence of

Figure 1.7 Models of tumour evolution and intra-tumour heterogeneity. (a) Clonal evolution model. High proliferation and genomic instability drive tumour expansion. Cells with the highest survival advantage are selected for resulting in tumours comprised of multiple clonal variants. (b) Cancer stem cell model. Heterogeneity arises due to mutations arising in stem cells or in progenitor cells at different stages of cell maturation. (c) Clonal evolution and the cancer stem cell model. Cancer stem cells can also undergo clonal evolution and so heterogeneity results from both clonal variants and mutations acquired in progenitor cells.

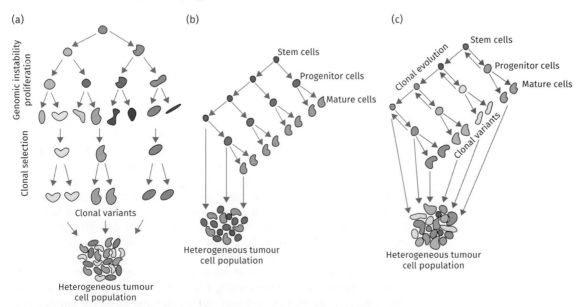

Adapted from Figures 2 and 3 in Fulawka et al. (2014). Biological Research, 47: 66–74.

Scientific approach panel 1.2
Evidence in support of the cancer stem cell model

The cancer stem cell model has been tested mainly using transplantation assays, which assess the potential of a cancer cell to form a tumour. Biopsies of human tumours are taken and the cells separated into phenotypically distinct subpopulations using flow cytometry based on cell surface markers expressed by the cells. These distinct populations are then transplanted into immunocompromised mice. If a tumour forms in the mouse, further rounds of cancer cell extraction, isolation of cell populations, and transplantations are carried out. Multiple isolations and transplantations yielding tumours are considered evidence of the existence of a cancer stem cell population within the tumour. One of the first cancers in which cancer stem cells were identified was in AML by Dominique Bonnett and John Dick in 1997. Through flow cytometry, cells were separated into minority and majority populations. The minority populations comprised approximately 1 per cent of the tumour population and expressed the cell surface proteins $CD34^+CD38^-$. These cells when transplanted into immunocompromised mice produced tumours and so were **tumourigenic**. In contrast the majority cell population that did not carry the same surface markers ($CD34^-D38^+$) had little or no capacity to seed a tumour. Furthermore, the $CD34^+CD38^-$ cells were able to give rise to tumours comprised of tumourigenic and non-tumourigenic cell populations in mice. Overall, the study provided evidence that AMLs were composed of small populations of self-renewing cells that are able to propagate tumours and a larger population of cells that do not demonstrate this capacity. Cancer stem cells have now been identified in a number of tumours including breast, colorectal, ovarian, pancreatic, and brain cancers. The cell surface markers on cancer stem cells vary by tumour type, for example the breast cancer initiating cells were found to be $CD44^+CD24^-$.

cancer stem cells experimentally is described in Scientific Approach Panel 1.2. Cancer stem cells play a role in drug resistance and in metastasis and thus challenge our ability to treat cancer. Both of these topics are discussed further in Chapter 6. See more about stem cells in Chapters 5 and 6

1.5 Viruses, inflammation, and cancer

Approximately 15 per cent of human cancers are known to be caused by viruses. Human oncogenic viruses include the human papilloma virus, the Epstein–Barr virus, human herpes virus type 8 (Kaposi), hepatitis B and C viruses, the human T-cell leukaemia virus, and the Merkel cell polyomavirus. Cancer-causing viruses can be DNA viruses or RNA viruses (retroviruses), of which some can integrate into the host genome upon infection of the host cell, whilst others are episomally maintained. Virus-induced cancers generally appear many years to decades after infection and not all infections with oncoviruses lead to cancer. This suggests that in addition to viral infection further oncogenic events for such as chronic inflammation, immunosuppression, or environmental mutagens are required for malignant transformation.

Viruses promote cancers through the production of protein products encoded by viral genes that disrupt key cellular processes implicated in cancer progression. Table 1.3 summarizes the common human viruses and the oncoproteins

Table 1.3 Oncogenic viruses and the major viral proteins by which they promote cancers

Virus	Type	Type(s) of cancer	Major viral oncoprotein(s)
Epstein–Barr virus (EBV)	DNA virus	Burkitt lymphoma, Hodgkin lymphoma, gastric and nasopharyngeal cancers	LMP-1, LMP-2A, EBNA-1, EBNA-2
Hepatitis B virus (HBV)	DNA virus	Hepatocellular (liver) cancer	HBX
Hepatitis C Virus (HCV)	RNA virus	Hepatocellular (liver) cancer	Core, NS3, NS4B, NS5A
Human papilloma virus (HPV)	DNA virus	Cervical, anal, oral cancers	E6, E7
Human T-cell leukaemia virus 1 (HTLV-1)	RNA virus	Adult T-cell leukaemia/ lymphoma (ATLL)	Tax, HBZ
Kaposi sarcoma-associated herpesvirus (KSHV)	DNA virus	Kaposi sarcoma	LANA, vFLIP, vGPCR, vCyclin
Merkel cell polyomavirus (MCV)	DNA virus	Merkel cell carcinomas	Large T antigen Small T antigen

by which they promote carcinogenesis. For example, a subset of the human papilloma viruses (HPV), the high-risk HPV types 16 and 18 that infect mucosal epithelial cells, are responsible for causing human cancers such as cervical cancer and some oral cancers. HPVs express two main viral gene products that are responsible for cancer formation, E6 and E7. E7 is able to bind to the tumour suppressor pRb and target it for degradation. This leads to the constitutive activation of the E2F transcription factors and uncontrolled entry into S-phase of the cell cycle. Unrestricted proliferation is detected by p53, and would normally lead to cell cycle arrest and/or apoptosis of the infected cell. However, the activity of p53 is also abrogated in HPV-infected cells through the E6 viral protein binding to and subsequently degrading p53. The actions of E6 and E7 are not limited to inhibiting p53 and pRb functions. E6 can also induce telomerase activity, contributing to cell immortalization whilst E7 can bind to an inactivated p21^{waf1} and p27^{kip1} cyclin dependent kinase inhibitors, thus preventing cell cycle inhibition.

Another mechanism by which viruses may promote cancer is by inactivating components of the DNA repair pathways. For example, the HBVX protein produced by the hepatitis B virus, associated with hepatocellular (liver) carcinoma, has been shown to inhibit the NER pathway, leading to genomic instability.

In some cases, cancers may be a consequence of the viral genome integrating into the host genome at locations that disrupt the activity of proto-oncogenes or tumour suppressor genes. One such example is the integration of the hepatitis B viral genome near the hTERT (human telomerase reverse transcriptase) gene. This leads to upregulated expression of telomerase, which contributes to cell immortalization.

In addition to these direct mechanisms, viruses can also promote cancer through an indirect mechanism. Persistent viral infection leads to chronic inflammation. Inflammatory cells (e.g. macrophages) can generate ROS, which in turn can cause oncogenic mutations in neighbouring epithelial cells. Inflammatory cells can also produce cytokines that activate pathways in pre-malignant cells driving cell proliferation and survival through signalling pathway involving NF-κB and STAT3. Chronic inflammation has been shown to play a prominent role in the development of hepatitis B virus induced hepatocellular cancer.

The role of inflammation in promoting cancer is not limited to viruses. Infection of the stomach lining with particular strains of the bacterium *Helicobactor pylori* (*H. pylori*) is associated with increased risk of stomach cancers. *CagA*-positive strains carry a gene called cytotoxin-associated gene A. To facilitate entry into the stomach cells, *H. pylori* injects the encoded toxin, CagA, into the stomach lining. CagA changes the structure of the stomach cells and allows the bacteria to attach more easily. Long-term exposure to the toxin induces chronic inflammation leading to stomach cancers in some infected individuals.

Chapter Summary

- Normal cells evolve to become cancer cells by acquiring successive mutations in primarily two classes of genes: the proto-oncogenes and the tumour suppressor genes.
- The protein products of these genes are involved in controlling a series of biological processes and therefore their alteration—genetic or epigenetic—leads to characteristic phenotypic changes described as the hallmarks of cancer.
- Only one copy of proto-oncogenes needs to be altered for oncogenic activity and therefore are autosomal dominant in their action.
- In contrast, tumour suppressor genes are recessive in that both copies of the allele are altered before tumour suppressor activity is lost. There are exceptions to this, with some tumour suppressor genes demonstrating haploinsufficiency.
- Mutations that drive cancer development and disease progression are called driver mutations; the rest are termed passenger mutations and their role in carcinogenesis is less clear.
- Between two and ten driver gene alterations are required to produce a transformed phenotype, although the cancer genome may carry hundreds of mutations.
- Our understanding of cancer is derived from *in vitro* cell culture studies, *in vivo* work in animal models, and using patient tissue or cells. Experimental work can involve single gene studies but increasingly exome or genome wide analysis.
- Large-scale cancer genome studies have provided deep insight into the mutational profile of the cancer genome, and is informing not only basic research but also the clinical management of cancer patients.
- The linear clonal evolution model explaining how cancers develop has now been replaced with much more complex branched models that better explain the heterogeneity observed within a tumour.

- This intra-tumour heterogeneity is caused by genomic instability, epigenetic changes driven by selective pressures exerted on tumour cells by the local tumour microenvironment and cancer stem cells.
- Cancer stem cells have the capacity to self-renew and differentiate into cell types that recreate the cellular heterogeneity of the tumour from which they derive.
- Viruses are associated with the development of some cancers as are certain bacteria.

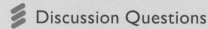 ## Further Reading

Friedberg, E. C. (2008). 'A brief history of the DNA repair field'. Cell Research 18: 3–7.
Provides a brief account of how the different DNA repair pathways were discovered.

Levine, A. J. and Oren, M. (2009). 'The first 30 years of p53: growing ever more complex'. Nature Reviews Cancer 9(10): 749–58.
Presents a succinct overview of different lines of evidence that led to the discovery of and establishment of p53 as a tumour suppressor, including its multiple functions.

Mesri, E. A., Feitelson, M. A., and Munger, K. (2014). 'Human viral oncogenesis: a cancer hallmarks analysis'. Cell Host & Microbe 15: 266–82.
Provides an overview of the molecular mechanisms that are disrupted by human oncoviruses and the cancer hallmarks that these disruptions contribute to.

Vogelstein, B., et al. (2013). 'Cancer genome landscapes'. Science 339(6127): 1546–58.
Reviews what has been learnt about cancer genomes from sequencing studies and what this information has taught us about cancer biology and future treatment strategies.

Vogt, P. K. (2012). 'Retroviral oncogenes: a historical primer'. Nature Reviews Cancer 12(9): 639–48.
Provides an overview of the discovery of retroviral oncogenes.

Discussion Questions

1.1 How are driver mutations identified and their functions validated?
Hint: think about cancer genome sequencing and follow on experiments.

1.2 What does the multi-stage model of cancer development mean? What is the evidence(s) for it?
Hint: think about the influence of age on cancer incidence, colorectal cancer histopathology, and the number of driver mutations in any given cancer.

1.3 What have viral cancers taught us about the process of tumour formation?
Hint: think about how oncogenes were discovered.

2 PATHOLOGY OF CANCER

Learning Objectives

- describe key pathological features of cancer;

- define the terms, tumour, neoplasm, benign, malignant, and metastasis;

- recognize the importance of differentiation, stage, and grade in the pathology of cancer;

- recognize and use the terminology for cancer types such as sarcoma and carcinoma;

- explain what the pathologist does in making a cancer diagnosis;

- discuss the significance of a tumour being more like an organ than a mass of cancer cells; and

- evaluate the complexity of cancer and be able to reduce it to a few underlying general principles—the hallmarks of cancer.

Cancer is a disease of dysfunctional communication within and between cancerous and adjacent non-cancerous cells. Understanding the way in which a sample of cancer tissue is a mixture of cancerous and non-cancerous cells is fundamental to research on cancer samples. It is also of prime importance to know the medical terminology used and something of the pathological features and complexities of cancer. Cancer affects the whole person since cancerous cells produce local and systemic signals, which modify normal cells and circulate throughout the body. It is the job of the pathologist to accurately identify, classify, and determine the severity of a cancer and to convey this information to clinicians caring for patients. This information will guide surgical interventions, chemotherapy options, and prognosis.

In this chapter, we discover how the pathologist carries out their part in the process of cancer diagnosis and prognosis and how this information can help in the choice of treatment. We will also see how understanding both the pathology and molecular biology of cancer is necessary for the derivation and comprehension of the hallmarks of cancer biology.

2.1 Benign, pre-malignant, and malignant solid tumours

As we know from chapter one, a neoplasm (from Greek meaning a 'new thing or shape') is a new growth of cells that results from their abnormal multiplication producing a tumour (from the Latin meaning a swelling). Tumours are behaviourally classified as benign or malignant based on characteristics identified by physician, pathologist, or radiologist. The gold standard diagnosis, however, is a pathologist-diagnosed sample of tissue (a biopsy), because other conditions, especially inflammation, can mimic tumour characteristics.

Benign tumours

Normal epithelial cells are tightly packed in a regular array behind a basement membrane. When cells outgrow their usual space but still remain behind a basement membrane or are encapsulated in fibrous connective tissue, they are called benign. Benign tumours are not classified as cancer because they lack the ability to invade adjacent or distant tissues. They grow by pushing adjacent structures aside. Examples of benign tumours are fibroids of the uterus (leiomyomas) or fatty lumps composed of lipocytes called lipomas that usually occur just under the skin. Benign tumours can still cause morbidity if they produce hormones or if they grow large enough to press on adjacent sensitive organs, e.g. pituitary tumours can produce excess hormones or affect vision by compressing the optic nerves.

Some causes of benign tumours include inherited mutations for example, in the neurofibromatosis type 1 gene (NF1). Mutations in NF1 are the most prevalent autosomal dominant disease, with an incidence of one in 3,000, and lead to benign tumours (neurofibromas) along with other symptoms. Benign tumours are usually easy to treat as they have defined boundaries facilitating complete surgical excision. Removal may be wise as occasionally malignant cells can arise in a benign tumour; for example, some intestinal polyps can become cancerous. Polyps with the potential to become cancer are then classified as pre-cancerous or pre-malignant rather than benign.

Pre-malignant lesions

Pre-malignant cells have microscopically and molecularly detectable changes that are associated with an increased risk of full-blown cancer developing. These morphological features are called cellular dysplasia (from the Greek meaning 'abnormal thing or form') or cellular atypia (from the Greek 'without form'). Such changes comprise microscopic features of abnormal size and/or shape termed 'pleomorphism' (from the Greek meaning 'different form'), nuclear envelope irregularities (pathologists talk about "nibbles" and 'indentations'), and increased hematoxylin staining of nuclei (hyperchromasia). An epithelial pre-malignant lesion has not invaded beyond the underlying basement membrane but may show sufficient dysplasia, and to such an extent that it warrants the diagnosis of 'carcinoma in-situ'. Thus there is a spectrum of pre-malignant dysplasia ranging from mild to moderate to severe, with the severe versions often referred to as carcinoma in-situ. The risk of cancer developing increases with the severity of dysplasia. Uterine cervical epithelial dysplasia is a good example of a pre-malignant lesion that can be detected by screening and thus eliminated at an early stage of disease. It is important to realize that not all pre-malignant lesions develop into cancer, and that regression of pre-malignant tissue is possible.

Malignant tumours

Malignant tumours arise either from pre-malignant lesions or *de novo*. The majority of these occur in epithelial sites, and although mesenchymal sites are much rarer, they are just as lethal. The best hallmark of an epithelial cancer is invasion of the malignant cells beyond the epithelial basement membrane. Malignant cells in the original or primary site can break down adjacent connective tissue and/or destroy adjacent normal cells to aid their invasion. Figure 2.1 shows a section through a mastectomy specimen harbouring an invasive breast cancer; the intrusive projections of the cancer give the appearance of a crab. When malignant cells reach a lymphatic or blood vessel, they can travel even further and form secondary tumours called metastases (singular-metastasis).

2.2 Nomenclature of solid tumours

In addition to benign or malignant, a further important classification that determines treatment and prognosis is the morphological classification. This is based on the type of malignant cell and its degree of differentiation. Differentiation refers to how well a cell has achieved its final function and form. For example, an epithelial stem cell will replicate but it won't perform the final function of producing mucin required from a glandular epithelial cell in the intestine, or produce keratin required for the function of a skin cell. The epithelial stem cell is therefore undifferentiated except to the point of being epithelial. Differentiation is identified by microscopic inspection or recognition of end products such as mucus or keratin. It's an old pathologist saying that the nucleus tells you whether a cell is benign or malignant, but the cytoplasm (containing the mucin or keratin) tells you what type of cell it is.

Figure 2.1 **An invasive carcinoma of the breast infiltrating normal breast tissue with a 'crab-like' appearance**.

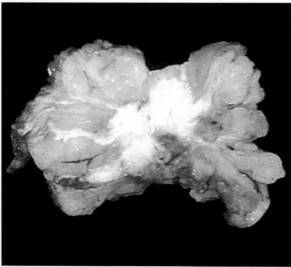

From http://www.pathologyoutlines.com/wick/breast/invasiveductalcarcinoma-gross55.jpg. Image contributed by Mark R. Wick MD.

Benign

Benign tumours are named based on their cell of origin with a suffix of '-oma'. Thus a benign tumour of bone is an osteoma and for fibrous tissue it is called a fibroma. If the benign tumour is composed of glandular tissue, then it is called an adenoma. Figure 2.2 (1) shows a benign adenoma that has not breached the basement membrane. A benign tumour with more than one type of differentiated cell is called a teratoma. Benign teratomas are encapsulated and do not invade surrounding tissues. A mature teratoma often contains different tissues including skin, bone, and teeth, none of which is associated with the area in which these structures usually occur. Teratomas form in many sites but especially in the ovary and testis.

Malignant

Malignant tumours of epithelial cells are given a suffix of 'carcinoma'; thus a malignant or cancerous tumour of glandular tissue is an adenocarcinoma. If the differentiation of the malignant cell is associated with connective or mesenchymal tissue, then the suffix is 'sarcoma'. As an example, for fatty tissue a benign tumour is a lipoma but a malignant tumour is a liposarcoma, similarly for bone a benign tumour is an osteoma and a malignant tumour of bone is an osteosarcoma. Anomalies in nomenclature have arisen due to common usage, such as melanoma, which is malignant with its benign counterpart being called a naevus or 'mole'. Look at Table 2.1 for these and other examples of how tumours are described.

Metastatic tumours

Metastasis is a major cause of cancer-related death and there has been little improvement in survival rates for patients who present with metastatic disease. Metastasis comprises two main stages: dissemination of malignant cells

Figure 2.2 **The multistep process of tumour growth from non-invasive to metastasis. BM = basement membrane; ECM = extracellular matrix.**

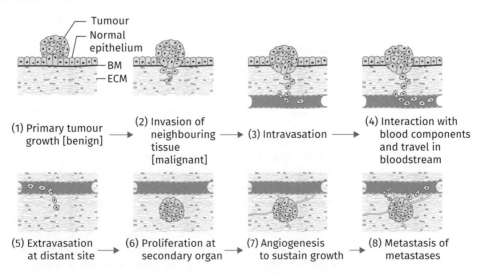

Figure 3.1 in https://clinicalgate.com/the-cellular-microenvironment-and-metastases-2, adapted from Figure 3 in Le, Q. T., Denko, N. C., and Giaccia, A. J. (2004). Cancer and Metastasis Reviews, 23: 293–310.

Table 2.1 Nomenclature of tumours

Tissue type	Benign tumour	Malignant tumour
Mesenchymal		
fibrous connective tissue	fibroma	fibrosarcoma
smooth muscle	leiomyoma	leiomyosarcoma
bone	osteoma	osteosarcoma
fat cell	lipoma	liposarcoma
Epithelial		
stratified epithelium	papilloma	carcinoma
glandular epithelium	adenoma	adenocarcinoma
Neural ectoderm	melanocytic naevus	melanoma
Germ cells	mature teratoma	teratocarcinoma

Tumours are named according to their cell of origin and whether they are benign or malignant. A suffix of –sarcoma delineates a mesenchymal origin whereas -carcinoma an epithelial one.

from the primary tumour to distant sites and, secondly, adaptation of these cells to grow successfully in the foreign microenvironment of a secondary organ. Failure of malignant cells to find a suitable niche may mean that although they can still become disseminated around the body they never successfully grow into full-blown metastases.

Many epithelial malignancies first spread from their original site into the tumour draining lymph nodes. It is now understood that malignant cells confer changes to the immune system that enable them to evade the action of immune cells and to destroy the normal architecture of the lymph node. The primary tumour produces signal molecules that communicate with the normal cells in nearby lymph nodes. Thus, signalling serves to remodel the lymph node tissue or stroma and dampen down immune responses thus providing a suitable niche for tumour growth.

To spread via the blood stream malignant cells have to overcome several hurdles. Look at Figure 2.2 (3–8) to see what these stages are. Cells must first enter the bloodstream through the lining or endothelial barrier of a blood vessel (intravasation), they must then survive in circulation before gaining access to a distant organ by again passing through the endothelial barrier of the blood vessel they are in. This is called extravasation. It is now thought that this last step involves malignant cells inducing cell death of the blood vessel endothelium in the target tissue. Favourable conditions for the growth of the extravasated tumour cells in this distant organ then lead to a metastasis. Metastatic spread involves epithelial cells taking on the properties of mesenchymal cells called epithelial to mesenchymal transition (EMT) and escaping into the blood stream. When the cells lodge in an organ to form a secondary tumour they often undergo the reverse process of mesenchymal to epithelial transition (MET). Therefore, a metastasis often has the histological appearance of the primary tumour. Examine Figure 2.3 that shows a section of a primary colorectal cancer and its lymph node metastasis, to see how similar the appearance of the primary and metastatic tumour looks.

> See more about metastasis in Chapter 6.

Figure 2.3 A haematoxylin and eosin (H&E) stained section of a well-differentiated primary colon cancer and a lower power view of its lymph node metastasis. Note in this case that the carcinoma retains its epithelial-like appearance in the lymph node metastases.

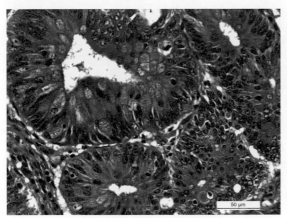 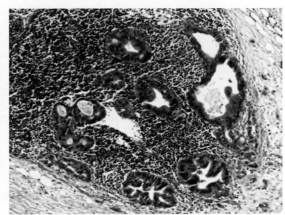

Primary colorectal cancer   Colorectal cancer in a lymph node

By kind permission of Dr NA Hung, Otago University, New Zealand.

Malignant cells can modify bone marrow derived cells that then prepare a premetastatic niche for the tumour in the distant organ. The target organ is often site-specific as malignant cells have a predilection for certain secondary organs. Stephen Paget first described this as his 'seed and soil' hypothesis. Liver, bone, and brain are common sites for secondary tumours, particularly melanomas and tumours of the breast and lung. Metastasis can pause before clinically significant secondary tumours are detected, a state known as dormancy that can last for years. We still don't know when or how a primary tumour seeds itself to form the cells that eventually become a metastasis. It is thought that these potential seeds may arise early in cancer development.

 Key Points

- Metastasis is the movement of tumour cells from their primary site to lymph nodes and to distant organs.
- It begins with invasion by a subclone of primary malignant cells into surrounding tissue, followed by intravasation into lymphatics and/or the blood stream.
- Invasive tumour cells undergo epithelial to mesenchymal transition (EMT), followed by mesenchymal to epithelial transition (MET) in a distant organ to form a metastasis.

2.3 Diagnosis of solid tumours

Solid tumours are distinct from the circulating tumours of blood cells (leukaemias) but even these so-called 'liquid' tumours sometimes form a solid mass.

There are various stages in diagnosis of solid tumours as outlined below starting with gross examination and then imaging, and finally histology.

Palpation and gross examination

A doctor examines the affected tissue, e.g. breast by gentle feeling, much like self-examination. This is called palpation. When a doctor examines benign tumours, e.g. breast lumps, they have some degree of mobility within the breast. This is because they are detached from the surrounding tissue due to the presence of a fibrous capsule. An example of a benign breast lesion is the fibroadenoma.

In general, benign tumours are uniform in appearance with no areas of hae-morrhage or cell death (necrosis), e.g. a mole or benign naevus. Malignant melanoma on the other hand has irregular margins with areas of bleeding and cell death.

Malignant tumours invade, and become embedded in the surrounding non-malignant tissue and are therefore non-mobile. Look at Figure 2.1 again to see how the tumour is anchored in the non-malignant breast tissue. Rapid and invasive tumour growth can lead to areas of necrosis or to bleeding due to blood vessel damage.

Scanning X rays and MRI

Imaging of a benign tumour shows a well-circumscribed edge whereas a rap-idly proliferating malignant tumour would show invasion of surrounding tissues, a loss of normal tissue architecture due to tissue destruction, and pos-sibly areas of necrosis where proliferation has outstripped supply of oxygen or nutrients. A mammogram is a low energy X-ray of the breast, which can detect cancer before it becomes detectable by palpation. For example, calcium deposits are easily detected by X-ray and can be indicative of the presence of cancer.

X-ray computed tomography (CT), the most common form of scan, is based on the ability of body structures to absorb X-ray energy. The resultant data generated from a large series of 2D images is processed by a computer to form a composite 3D image that allows the user to 'see' virtual slices through the subject. Differences in tissue density as little as 1 per cent can be dis-tinguished. Contrast CTs are produced by injection of a radiological contrast medium, e.g. gadolinium, prior to scanning. This highlights structures such as blood vessels that are not easily seen in a non-enhanced CT scan taken with-out contrast. See Figure 2.4 for an enhanced CT image of a breast carcinoma. Positron-emission tomography is another type of computed or CT scan called a PET scan; it involves injection of a radio-labelled metabolic precursor (tracer) such as fludeoxyglucose, an analogue of glucose that can then be traced. This procedure although expensive is particularly suited to detecting metastases that have a high uptake of the tracer due to their elevated glucose require-ments. A PET scan of a breast carcinoma is also shown in Figure 2.4. MRI on the other hand uses magnets instead of X-rays so avoiding radiation effects. It is invaluable for detecting cancers of unknown origin when the patient pres-ents with general symptoms indicative of cancer such as sudden and unex-plained weight loss or pain. Whole body MRI scanning may be used to detect metastatic tumours. Suspicious features are then followed by a biopsy and histological examination.

Figure 2.4 **Scans of cancer in the breast.** On the left is a CT scan showing a breast tumour highlighted; on the right is a PET scan of a breast tumour.

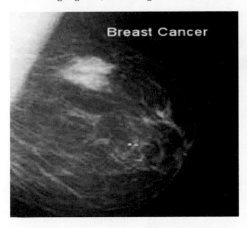 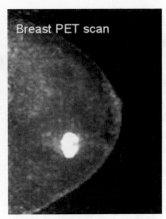

LHS image: breast cancer taken from https://breast-cancer.ca/oncolgst; RHS image taken from https://breast-cancer.ca/pets-bcanc.

Biopsy and histological examination

The pathologist carries out preparation and microscopic examination of affected tissue, usually from a surgically removed biopsy or resection, but possibly from a fine needle tissue aspirate. Once dissected but prior to looking at tissue under the microscope, it is processed and stained to show up features such as nuclei and also expressed proteins of interest for a diagnosis.

Staining sections for histopathology

Tissues are first fixed to preserve their structure and then thin sections or slices are cut and mounted on glass slides. Tissue is then stained using hematoxylin and eosin (H&E). Hematoxylin stains nuclei blue and eosin stains cytoplasm and connective tissue pink/red. Microscopically, cells of benign tumours have a close resemblance to their normal counterparts, in other words they are well differentiated. Malignant tumours on the other hand invade and destroy surrounding tissue and have an irregular infiltrative margin. They have evidence of high proliferation (mitotic figures or expression of proliferation markers such as Ki67) and necrosis or cell death. Normal tissue architecture is destroyed and blood vessels appear poorly formed.

Immunohistochemistry is used to detect the location of specific proteins in a section. Antibodies directed to the whole or part of the particular protein are applied to a tissue section that has been rehydrated as antibodies only react in the aqueous phase. Bound antibody marks the location of the protein of interest and this is then detected using a variety of enhancing and staining products, for example immunofluorescent or chromogenic substrates. A hematoxylin counterstain is then applied before dehydrating and cover slipping the section for microscopy. Immunostaining in tumour diagnosis uses tissue specific markers as shown in Table 2.2. Figure 2.5 shows how immunostaining for the mutant form of p53 stains the malignant gastric carcinoma cell nuclei brown. RNA can

Figure 2.5 Immunostaining of mutant p53. (a) The highly stable mutant p53 is strongly stained brown in the nuclei of the tumour cells of a gastric adenocarcinoma. (b) A schematic of the process of immunohistochemistry.

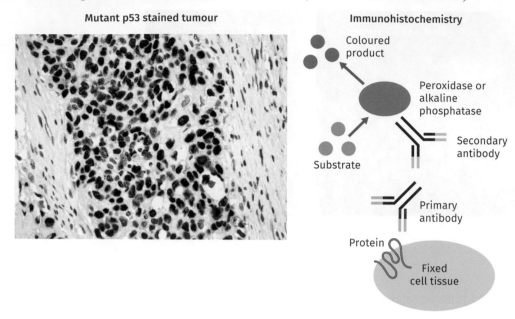

(a) From Abcam ab 32049: https://www.abcam.com/mutant-p53-antibodyy5-ab32049.html. (b) Adapted from Rockland: https://rockland-inc.com/ihc-products.aspx.

Table 2.2 Tissue specific markers

Marked cells	Protein	Function
Astrocytes	Glial fibrillary acid protein (GFAP)	Intermediate filament
Breast	HER2, ER, PR	Receptor tyrosine kinase
Epithelial	Cytokeratin	Intermediate filament
Leukocyte	CD45	Cell surface receptor
Lymphoma	Bcl2	Anti-apoptotic protein
Mesenchymal	Vimentin	Intermediate filament
Prostate	Prostate specific acid protein (PSA)	Seminal plasma protein
Vascular endothelial cells	Tie2/Tek	Receptor tyrosine kinase
	VEGFR	

Certain proteins are highly expressed in a tissue specific manner and this is used for their identification in immunohistochemistry.

❯ See more about p53 in Chapter 4.

also be detected in tissue sections using *in-situ* hybridization and an oligo-nucleotide probe in the place of antibodies.

Liquid biopsy

When cancer cells are shed into the blood stream, their detection there can be used to monitor cancer progression and response to treatment. This technique, known as 'liquid biopsy', holds the promise of detecting the

Figure 2.6 Liquid biopsy. Possible applications of a liquid biopsy of a lung tumour. Blood-based biomarkers can be obtained from circulating tumour cells and nucleic acids released from dead tumour cells.

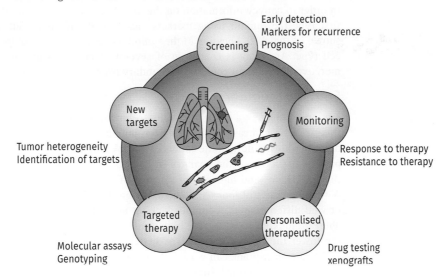

Figure 1 in Zhang, Z., et al. (2015). Frontiers in Oncology, (5) article 209. doi: https://doi.org/10.3389/fonc.2015.00209.

appearance of metastatic cancer before it becomes clinically overt. Liquid biopsy can replace the tissue biopsy if the tumour is in an inaccessible location where sampling could be hazardous. Diagnosis can be made by identifying circulating tumour cells (CTC) using cytometric approaches such as immunohistochemistry or fluorescence microscopy for protein markers. An initial enrichment step such as flow cytometry may be required if the CTCs are present at low amounts. The presence of clusters of contiguous cells (micro-emboli) that might have an increased capability to go on to form metastases could be prognostic. CTC detection can also be used to monitor the response by the cancer to therapeutic interventions so that optimum treatment can be given. Monitoring cancer spread is medically challenging and at present relies mainly on radiographical methods such as MRI but tumours need to grow to a sufficient size before these techniques can detect them.

When cells die, they release DNA into the blood stream so live CTCs are not always required for this technique. Tumour DNA is different from normal cellular DNA and these differences, such as gene mutations, can be used to detect circulating cancer DNA (ctDNA) load in the blood stream or other body fluids such as cerebrospinal fluid (CSF) for brain cancer. Reading the genetic signature of the tumour DNA using next generation sequencing can give diagnostic information, with the added benefit of being minimally invasive. The possibility of detecting tumour specific RNA, particularly that is bound to extra cellular vesicles in body fluids, is also being researched. Figure 2.6 shows the diversity of ways in which liquid biopsy can be used for diagnosing and monitoring cancers.

❯ See more about liquid biopsy in Chapter 6.

2.4 Tissue typing, grading, and staging

In order to convey information on the broad nature of a tumour, doctors must first type the tumour using markers based on the putative cell of origin as shown in Table 2.2. Location of the tumour is not a guarantee of its originating cell type. An organ is composed of several potentially cancerous cell types; moreover, the tumour could be a secondary from an unknown primary tumour.

Epithelial tumours

Tumours are given a grade according to how well they are differentiated, and this grade is a measure of aggressive potential related to prognosis. Grade describes how abnormal the cancer cells look compared with normal cells. The pathologist looks at the changes in both the organization of the tissue and the size, shape, and intensity of staining of the cells and their nuclei by H&E. A tumour that still looks similar to normal healthy cells is called low grade or well differentiated. A tumour lacking normal tissue architecture and with aberrant looking cells is called high grade, or poorly differentiated. In general, the lower the tumour's grade, the better the prognosis. The factors used to assign a cancer grade can vary with the type of cancer, but in general, grades from 1 to 4 are ascribed. Grade 1 is well differentiated and grade 4 is undifferentiated, based on the degree of abnormality and deviation from the normal counterpart.

Stage is a measure of how far the tumour has spread and this includes its size. Staging is prognostic and important for determining treatment options. It states whether cancer cells have spread to local tissues or to the lymph nodes located near the tumour. The pathologist will look for tumour margins in a surgically resected tumour to determine if all visible tumour has been removed. If possible, the pathologist will state whether the tumour has spread to other parts of the body and metastasized. A cancer is always reported as the stage it was at first diagnosis even if the cancer spreads later. The TNM classification takes the size of the tumour (T) into account, unless it has metastasized. The number following N represents the number of lymph nodes containing histologically visible cancer (lymph nodes involved) or following the M is the number of metastases found. The TNM system of staging is used for most types of non-haematological cancers except brain cancer, which rarely spreads outside the CNS. Figure 2.7 shows how colorectal cancer (CRC) progresses and how it is staged. Globally CRC is the third most common cancer and occurs particularly in older people but there is an alarming trend of increased incidence in a younger population. Colorectal carcinoma originates from epithelial cells lining the gut that acquire defects in their ability to control proliferation and differentiation. Stage '0' is *in-situ* and lies behind the basement membrane (BM) on the mucosa. Stage 1 has invaded beyond the BM into the muscle layer of the bowel wall but does not involve any lymph nodes. In stage 3, the tumour has breached the serosal wall of the bowel and now occupies space in the abdominal cavity. Note that in this case the stage 3 tumour has also invaded local lymph nodes. Stage 4 CRC has spread to further lymph nodes and to distant organs. Table 2.3 shows the Tumour, Node, Metastasis (TNM) system used for staging carcinomas. The Dukes' staging scheme named after Dr Dukes, as shown alongside, is sometimes used for colorectal cancer.

Figure 2.7 Staging colorectal cancer. Stage 1: The polyp remains inside the mucosal layer, facing the lumen of the bowel; Stage 2: The cancer has spread into the muscle layer; Stage 3: The cancer has spread to lymph nodes; Stage 4: The cancer has spread to other organs.

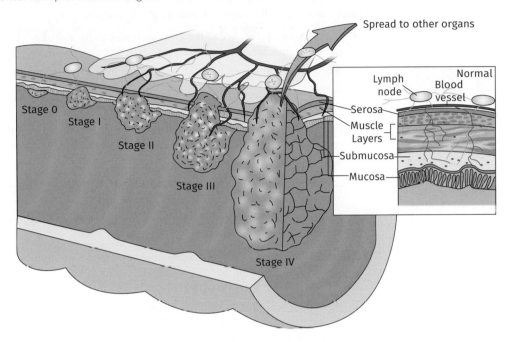

From Terese Winslow, National Cancer Institute. Colon cancer staging: https://www.teresewinslow.com/#/digestion. As shown on Ant Yradar. Advances in Colorectal Cancer research: http://antyradar.info/stages-of-colorectal-cancer/stages-of-colorectal-cancer-bowel-cancer-staging-size-position-spread.

Table 2.3 Staging cancer

TNM classification (American joint commission on cancer)				Dukes' classification for colorectal cancer
Stage	T	N	M	Stages
Stage 0	Tumour *in-situ*	N0	M0	
Stage I	T1	N0	M0	A
	T2	N0	M0	B1
Stage II	T3	N0	M0	B2
	T4	N0	M0	B2
Stage III	T1 T2	N1 or more	M0	C1
	T3 T4	N1 or more	M0	C2
Stage IV	Any T	Any N	M1	D

The stage of a disease represents how far it has spread and the main classification system in use is the TMN. The Dukes' system is sometimes quoted for colorectal cancer.

The pathology report

A pathology report gives a diagnosis based on the pathologist's examination of tumour tissue and will contain the following information:

- macroscopic appearance of the gross specimen to the naked eye such as size of the tumour, which is confirmed microscopically;
- the type and sub-type of cancer, such as carcinoma or sarcoma;
- tumour grade;
- lymph node status—do the lymph nodes contain tumour cells?;
- status of the tumour margin—are there cancer cells at the edge of the resected tissue suggesting that not all the cancer has been removed?;
- presence of metastases;
- results of immunohistochemistry and molecular tests for expression of tumour markers, e.g. HER2 for breast cancer; and
- a measure of how proliferative the tumour is e.g. mitotic count or degree of staining for a proliferation marker, e.g. Ki67.

This report will be used by the oncologist to identify treatment options.

Cancer of blood and lymphatic cells

Thus far, we have looked at solid tumours that arise in epithelial or mesenchymal tissues. Cancer can also affect cells in the blood, bone marrow, and lymph nodes. Leukaemia is a group of cancers of white blood cells (granulocytes/myeloid cells), lymphoma is a group of cancers of lymphoid cells such as B and T lymphocytes, and myeloma is a cancer of the plasma cells that normally produce antibodies. These cancers are classified on the basis of cell lineage and degree of differentiation of the malignant cells. The myeloid lineage gives rise to myelogenous or myeloid leukaemias, e.g. acute myeloid leukaemia (AML) and chronic myeloid leukaemia (CML). The lymphoid lineage gives rise to acute lymphoblastic leukaemia (ALL) and chronic lymphocytic leukaemia (CLL), as seen in Figure 2.8.

Lymphomas are historically divided into two main categories: Hodgkin and non-Hodgkin, depending on the kind of lymphocyte involved. Hodgkin

Figure 2.8 Classification of leukaemias. Leukaemias are cancers of the blood derived from lymphoid and myeloid stem/precursor cells. Leukaemias are further classified according to whether they are chronic (slow growing) or acute (fast growing). The subtypes are: acute myeloid leukaemia (AML), chronic myeloid leukaemia (CML), acute lymphoblastic leukaemia (ALL), and chronic lymphocytic leukaemia (CLL).

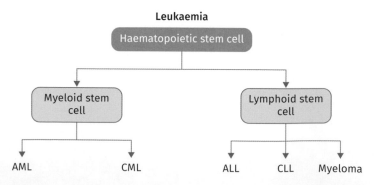

lymphoma is more common in younger patients and usually carries a high survival rate. It is characterized by a giant cell called the 'Reed-Sternberg cell'. Non-Hodgkin lymphoma (NHL) is a group of cancers involving malignant lymphocytes that are more commonly diagnosed in patients over sixty-five years. NHL can be aggressive or indolent and diagnosis is by surface markers to distinguish their differentiation. The most common NHL is B cell lymphoma of which diffuse large B cell lymphoma and follicular lymphoma are the most prevalent. Burkitt lymphoma is a rare and aggressive form of the disease associated with Epstein–Barr viral infection. T cell lymphoma and natural killer (NK)-cell lymphoma are relatively rare, although T cell lymphoma is more frequent in Asia.

Some of these myeloid or lymphatic cancers are underpinned by characteristic chromosomal translocations and the corresponding effect on proliferation drivers, such as Myc or anti-apoptotic molecules such as Bcl-2.

Biomarkers and cancer

Molecular pathology is being increasingly introduced into routine assessments and is complementary to histopathology in aiding diagnosis, prognosis, and to predict response to therapies (predictive). Molecules that are indicative of the presence of a disease such as cancer are referred to as biomarkers.

A tumour biomarker maybe a molecular change in DNA, RNA, protein, or a metabolite produced by cancer cells or by other cells in response to the cancer. Tumour biomarkers can be detected and/or monitored in tissue, blood, or in urine, stool, or sputum. Various tests can be used to identify a biomarker including immunohistochemistry or immunofluorescence to detect protein levels, DNA sequencing or fluorescence in-situ hybridization (FISH) to identify genetic changes and microarrays or qRT-PCR to identify changes in gene expression.

Diagnostic markers provide evidence of a type of cancer. For example, elevated prostate specific antigen (PSA) in blood may indicate the presence of cancer in the prostate. BRAF mutations in melanocytes are indicative of a melanoma and the presence of BCR-ABL is diagnostic for CML. Diagnostic biomarkers can also be prognostic, e.g. PSA is used to inform about disease recurrence or disease progression. Prognostic markers are used to stratify patients at high or low risk of disease recurrence and therefore identify patients who should or should not receive adjuvant treatment. In contrast, predictive biomarkers are used to assess whether an individual will respond favourably or not to a particular treatment. Thus, if the decision is to administer adjuvant therapy, then predictive markers can guide in the selection of the most suitable therapy or combination of therapies for a given patient.

Mutant forms of some proteins such as IDH1, EGFR, and HER2 can be assessed using IHC and may be of diagnostic, prognostic, or predictive value. For example, the presence of mutant EGFR indicates not only malignancy but also possible tumour response to tyrosine kinase inhibitors whereas mutant Ras that acts downstream of EGFR, indicates resistance to EGFR inhibitors. Another example is the presence of HER2 in breast cancer, which is predictive for herceptin treatment, and ER+ for anti-oestrogen therapy. The field of biomarkers is expanding rapidly and is covered in Chapter 6.

❯ See more about biomarkers in Chapter 6.

Artificial intelligence

Artificial intelligence (AI) and machine learning technology are being applied to histopathological images of tumours with the aim of providing computer-aided diagnosis. Machine learning-derived sets of rules or algorithms based on histopathological features and validated by pathologists can be used to classify cancers accurately. Nicolas Coudray and colleagues in 2018 developed such a system to distinguish adenocarcinoma from squamous carcinoma of the lung. From the pathology images, the model was also able to predict six of the most common mutations found in lung cancer, including *KRAS*, *EGFR*, and *TP53*. This has great implications for determining the best treatment for the patients with adenocarcinoma and squamous carcinoma of the lung and give future impetus to finding new drugs to target specific mutations. This technology will aid with diagnosis but the pathologist will carry out staging and monitor response to treatment. The use of AI in the diagnosis of tumours is likely to increase in future.

The ethics of working with human tissue

The use of human tissue and clinical data for research is of paramount importance for understanding disease mechanisms. All collection, storage, and use of such material require written ethical approval from an ethics committee. The committee is charged with ensuring that proposed research is valid and carried out in accordance with statutory regulations: for example, in the UK this is governed by the Human Tissue Act. Patients must give informed consent for their tissue and clinical details to be used for research. All tissue and data must be anonymized and stored securely.

2.5 Cancer screening

A pre-cancerous lesion or condition is one in which there is a statistically significant possibility that it will undergo progression to malignancy. Therefore, if pre-malignant lesions can be diagnosed and treated easily, the risk of developing cancer can be reduced. This is the rationale behind cancer screening.

Cancer screening involves looking for cancer before a person has any symptoms, with the aim of catching cancer at an early stage when it is easier to treat. To be of value screening must target a common form of cancer for which available treatment is capable of reducing death rates. Tests must be of high predictive value so that false results are minimized. False positives cause unnecessary stress for the individual, particularly as follow-up often involves expensive and invasive procedures. False negatives give unrealistic hope and deny the individual the investigations they require for a speedy true diagnosis and early intervention. Detected early cancers may result in overtreatment with a concomitant morbidity that can even reduce life expectancy due to drug side effects. For example, radiotherapy and chemotherapy for breast lesions can increase the risk of breathing problems or even heart disease. Some cancers do not cause symptoms and never become life threatening so detecting such a cancerous or precancerous lesion may involve the patient in unnecessary procedures that do not improve or extend life. Some cancers are not suitable for screenings as they arise *de novo* with no precursor lesion, and others are preceded by stages of incipient malignancy that are undetectable. Moreover, not all precancerous lesions progress to malignancy. Interestingly a recent study showed that a single screening test by sigmoidoscopy in people over the age of fifty years reduces the incidence of bowel cancer for men, but it had little or no effect on outcome for women. Screening may involve testing body fluids such as blood or urine, tissue sampling such as a cervical smear, or imaging such as in breast screening. Current screening is carried out for cervical, breast, colorectal, and prostate cancers. Any screen detected potential cancer is investigated further and diagnosis is usually made from tissue biopsy samples, magnetic resonance imaging (MRI) or computed tomography scan (CT scan), and blood tests.

Screening mammograms are given for women who have no symptoms. If cancer is suspected, a more detailed diagnostic mammogram can then be given. Studies have shown that finding and treating screening detected cancers can reduce mortality from breast cancer in women between the ages of fifty and seventy. These benefits have to be weighed against harms, which include false positives with the associated unnecessary surgery and anxiety, false negatives, and radiation exposure during screening. However, adverse effects of radiation are thought to be minimal, as mammography screening every two to three years does not appear to affect overall cancer mortality including breast cancer. One in 2,000 women will have their life prolonged if they undergo ten years of screening, but in this time ten healthy women will have unnecessary treatment. Targeted screening that is genetically or epidemiologically informed has been proposed that will benefit people at greatest risk. This should remove costs, both personal and financial, of over diagnosis and other negative consequences of mass screening. For example, for women with a known risk factor for breast cancer such as mutations for the *BRCA2* gene or family history, mammography is important. A breast MRI in addition to mammography may be recommended for women who are at high risk of breast cancer, but it is not recommended as a screening tool as it may miss too many tumours.

Case study 2.1
Lessons from cervical cancer

A. Cervical smear screening

Cervical cancers are usually squamous cell carcinomas of the epithelial cells that cover the cervix. In the UK, it is estimated that one in 130 women will be diagnosed with cervical cancer during their lifetime, with incidence rates projected to rise. Risk factors for cervical cancer include age at first intercourse (<17 years significant), sexually transmitted diseases and human papilloma virus (HPV) in particular, smoking and components of seminal fluid. For example, prostaglandin E2 (PGE2) present in seminal plasma may play a role in modulating neoplastic cell function and promoting cervical tumorigenesis.

Cervical smears remove some of the surface epithelial cells from the cervix for staining and microscopic examination. Pathologists look for atypical changes to these cells and classify cervical lesions by their **cytology**. Mild, moderate, and severe changes in the shape of the cells and their nuclei such as increased nuclear to cytoplasmic ratio and irregular chromatin pattern are noted. This can be given a grade of 1 to 3 for the extent of the cervical intra-epithelial neoplasia or CIN present. An alternative assessment scheme is the Bethesda Classification, which grades a squamous intraepithelial lesion (SIL) as low or high grade. The most severe classes of atypia or dysplasia are sometimes called 'carcinoma *in-situ*' (even though this is a contradiction in terms!). Perhaps the better term to describe these atypical cells is **dyskaryosis**. Look at Figure CS 2.1 that shows the comparison between normal cervical epithelial cells with small nuclei in comparison to cytoplasmic volume and cells of increasing dysplasia. A biopsy is necessary to determine the degree of invasion of the altered cells.

B. Biopsy

Pre-neoplastic changes are classified on the proportion of the cervical epithelium that has atypical cells and this requires taking a biopsy of the cervix and histological examination of stained sections. If left untreated, CIN can go on to become cancerous; however, the majority of women with CIN do not get cancer. Fifty per cent of CIN1 spontaneously regress but 20 per cent can progress over ten years to CIN2 or CIN3. Twenty per cent of CIN3 patients develop carcinoma of the cervix over the next ten years and this rises to 40 per cent over twenty years. Thus, for the population as a whole, within approximately twenty years 20 per cent of 20 per cent (i.e. 4 per cent) CIN 1 patients would develop carcinoma if untreated. However, some individuals can progress from CIN 1 to carcinoma in a matter of months rather than years.

As cervical cancer risk is associated with the presence of human papilloma viral infection, the biopsy tissue can be stained for the presence of HPV. Infection shows as squamous cells with large cytoplasmic vacuoles called koilocytes. HPV infection with high-risk types (HPVs 16, 18, 31, 33, 35) is a risk factor for the development of cancer. To specify the type of HPV present requires *in-situ* hybridization or PCR for the viral genome. Vaccination against some of these high-risk HPV types is available in several countries.

Figure CS 2.1 Cervical cells on PAP smear. Cellular changes seen in a series of Papanicolaou (PAP) stained cervical smears of increasing levels of dysplasia and their related appearance in the cervical epithelium.

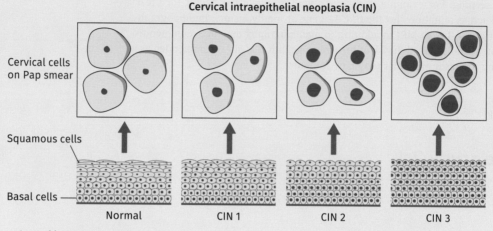

Taken from https://www.123rf.com/photo_11578741_cervical-dysplasia-stages-and-pap-smear-test.html.

2.6 Hallmarks of cancer

Knowledge gained from the study of cancer pathology, along with cellular and molecular data has led to the current understanding of the nature and distinguishing features or hallmarks of cancer. Douglas Hanahan and Robert Weinberg first coined the term 'hallmarks of cancer' in 2000 to describe the essential biological processes necessary for tumour formation.

These hallmarks are generalized features that are considered fundamental to cancer cell biology. However, strictly speaking, 'cancer' is a term for invasive malignancy, and many of these hallmarks also apply to benign neoplasms.

It is now recognized that the hallmarks of cancer must be taken in context of both malignant cells and recruited non-malignant cells, as both contribute to tumour development and progression. A tumour is not a conglomeration of proliferating cancer cells but the result of complex interactions between the malignant and non-malignant cells in the tumour microenvironment. Cancer cell lines, popular tools of the molecular biologist, are not fully representative of cancer cells *in vivo,* where tumour cell heterogeneity and cell-cell interactions are key features. The non-malignant cells of the microenvironment are not passive normal bystanders but are 'hijacked' by the malignant cells into providing supportive roles. The contributions of recruited stromal cells and the cancer niche are now included in the updated 2011 hallmarks of cancer. These ten hallmarks are:

1. unrestricted proliferation due to deregulation of growth signals and failure of their negative controls;

2. sustained growth of cancer cells by avoiding growth suppression signals that in normal cells are hard-wired into the control of cell numbers;

3. evasion of cell death mechanisms by inhibiting apoptosis or autophagy;

4. generation of an efficient blood supply including new blood vessel formation via the process of neo-angiogenesis;

5. the ability to overcome the normal controls that maintain cell polarity, cell-cell and cell-matrix contact. This enables invasion by cancer cells into adjacent tissues and ultimately may lead to metastasis;

6. development of a mutator phenotype and an increased tolerance of DNA mutations. Epigenetic changes can also afford a rapid and global field change in a cell and may underpin several forms of carcinogenesis;

7. gain of replicative immortality by inducing a telomere maintenance mechanism to prevent crisis due to critically short telomeres;

8. increased DNA replication and cell division requires energy and cancer cells often revert to a form of glycolysis to boost their supply;

9. for their survival, cancer cells must evade the immune response that should be mounted to remove these altered cells; and

10. development of a supportive stromal microenvironment often with tumour-promoting inflammatory conditions.

We will now discuss each of these in more detail.

Proliferation

A fundamental trait of cancer cells is an ability to drive chronic proliferation. Normal adult cells control cell division to provide only the cells that are required for renewal. It is a misconception, however, that cancer cells always divide faster than their normal counterparts. Normal cells lining the gut are replaced at a higher rate than most colorectal tumour cells. Thus, treatments that kill dividing cells could adversely affect the gut lining, leaving the tumour relatively intact.

Cancer cells acquire the ability to proliferate in various ways including deregulated production of growth factors e.g. insulin-like growth factors such as IGF1 or mutant receptors such as mutant EGFR that no longer require ligand for activation. All can lead to constitutive activation of proliferation or proliferative independence. It is important to understand that excessive stimulation of cell division does not on its own result in cancer development. Inappropriate proliferative signals actually provoke counteracting responses such as senescence or apoptosis. Thus, in many cancers the actual levels of mitogenic stimulation probably reach a compromise between proliferation and induction of such anti-proliferative defence mechanisms. Higher levels of proliferation

❯ See more about prolif-
eration pathways in Chapter 4.

may be achieved if senescence and apoptosis pathways are disabled or at least uncoupled from excessive cell division.

Evasion of growth suppression

Cancer cell proliferation necessitates the circumvention of pathways that negatively regulate cell proliferation. Mutation or deletions of the tumour suppressor genes that negatively regulate cell division are common hallmarks of cancer. The 'belt and braces' of this process falls mainly to the two powerful tumour suppressor genes, *RB1* (retinoblastoma associated gene) and *TP53*. *TP53* has been dubbed 'the Guardian of the Genome' for its central role in homeostasis.

The RB pathway acts as a gatekeeper for cell cycle progression by binding to and inhibiting the transcription factor E2F required to initiate DNA replication and cell cycle progression. This can be seen diagrammatically in Figure 1.3. p53, on the other hand, responds to cellular stress, detecting DNA damage or insufficiency of building blocks and energy for complete and accurate replication of genomic DNA and adjusting the rate of proliferation or senescence accordingly. If the alarm signals are very intense, wtp53 is stabilized and induces cell death by apoptosis. Many tumours harbour a mutant form of p53 and this is often far more highly expressed than wtp53, probably due to its increased stability. This fact is used to detect likely mutant p53 is tissue sections by immunohistochemistry. Figure 2.5 shows how immunostaining for p53 highlights cells harbouring a stable mutant.

❯ See more about p53 in
Chapter 4.

Evasion or hijacking of cell death mechanisms

Apoptosis is crucial for normal embryonic development, cell turnover, and immune effector functions. Importantly apoptosis is a mechanism for cell death. Apoptosis was first described in the 1970s by John Kerr and Andrew Wylie who defined characteristic death associated changes in cell morphology, including cell shrinkage, chromosome condensation, and nuclear blebbing.

The apoptotic machinery comprises two major components: apoptotic signalling and apoptotic execution. The regulators of apoptosis in turn are divided into upstream ones that receive and integrate extracellular death signals (e.g. Fas) and those that regulate the downstream intracellular effector programme (e.g. p53). The triggers that convey signals between regulators and effectors and put the cell into irreversible destruction are members of the Bcl-2 family. There are pro- and anti-apoptotic family members and the balance between the two determines the outcome for the cell.

Tumour cells have evolved a variety of strategies to avoid apoptosis, including loss of *TP53* function or increased expression of anti-apoptotic regulators such as Bcl-2 relative to the pro-apoptotic members such as Bax or Bim. The signalling pathways associated with apoptosis are covered in Chapter 4. There are many ways of determining apoptosis in tissues. Firstly, the nucleus breaks into fragmented or apoptotic bodies that are seen in H&E stained sections and reported by the histopathologist. This can be confirmed by staining for cytochrome C that is released from the mitochondria into the cytoplasm during apoptosis. Chromatin condensation and DNA fragmentation can be detected using terminal deoxynucleotidyl transferase dUTP nick and labelling (known as TUNEL). Apoptosis can be measured in cell suspensions by flow cytometry using makers such as anexin V that are restricted to the inner cellular membrane in intact viable cells but which become externalized during apoptosis.

In contrast to apoptosis, which removes cells through phagocytosis by others, necrosis is a death process involving damaged cells releasing their contents into the surrounding tissues. Although for many years necrosis was considered an uncontrolled form of cell death it is now recognized that it is under genetic control and not a random process. Distinguishing necrotic cells is usually done by morphological analysis in H&E stained tissue sections where necrotic material lacks cellular organization and staining. Immunohistochemistry using antibody to RIPK1 (receptor interacting serine/threonine kinase 1) can be used to confirm but since necrotic cells lack intact protein or nucleic acid this is often not ideal.

Autophagy is now recognized as another important pathway for programmed cell death, and is particularly induced in conditions of nutrient deficiency. The autophagic pathway leads to break down of cellular organelles allowing the resultant catabolites to be reused in energy metabolism or biosynthesis. Mice made deficient in the autophagy pathway are susceptible to cancer suggesting autophagy can be tumour suppressive. Autophagy is studied in tumour sections by immunohistochemistry for protein expression from the autophagy related genes (ATGs), e.g. beclin1 (*ATG6*) and also LC3A/B (*MAP1A/MAP1B*), involved in the formation of autophagosomes. This is covered in more detail in Chapter 4. Paradoxically radiotherapy and some chemotherapeutic drugs can induce high levels of autophagy that act in a cytoprotective manner, providing building blocks for new cancer cells. In this way, autophagy is working as a form of drug resistance. These drug resistant cells can survive and later form the basis of regrowth of the tumour.

Necrosis, unlike apoptosis or autophagy releases pro-inflammatory signals into the microenvironment so rather than avoiding this form of death tumour cells hijack the resultant immune inflammatory cells into roles of tumour promotion. These specialized immune cells can promote angiogenesis and as such aid cancer cell proliferation and invasion.

> See more about cell death pathways in Chapter 4.

Neo-angiogenesis

Proliferation and survival of cancer cells requires a blood supply to provide nutrients and oxygen and remove metabolic waste. Tumours' needs are partially met by acquiring new vessels through the process of neo-angiogenesis. Pro-angiogenic signalling coupled with inhibition of the normal quiescent state of the vasculature support the expansion of neoplastic growth. A proto-type of an angiogenic stimulator is vascular endothelial growth factor (VEGF-A) whereas a major inhibitor is thrombospondin (TSP-1). VEGF signalling is up regulated by oncogene activation, hypoxia, or indirectly by tumour associated immune inflammatory cells and can be stained for in tissue sections. TSP-1 is a key counterbalance that is regulated by p53. Tumour vasculature is characteristically abnormal causing erratic blood flow, transient hypoxia, with leakiness and sometimes haemorrhages. High blood vessel density and the present of haemorrhage are diagnostic of a high grade in many cancers such as glioma. Neo-angiogenesis is an important part not only of invasion but also of the metastatic process, providing as it does a blood supply for tumour cells arriving in a distant organ.

Activation of invasion and metastasis

Normal epithelial cells are adherent to each other and to their basement membrane matrix. This provides a brake on unsolicited proliferation and invasion. Loss of cell-cell contact and cell-extracellular matrix (ECM) adherence are key hallmarks of cancer cells. This allows cells to escape from the inhibition of proliferation and invasion that is provided by cell contacts. Tumour suppressor genes such as the neurofibromatosis gene *NF2* orchestrate contact inhibition. The *NF2* gene product Merlin achieves this by regulating cytoskeletal dynamics, strengthening cell-cell contacts, and sequestering growth factor receptors to maintain homeostasis in architecturally complex tissues.

Adhesion molecules that function during embryogenesis or during inflammation are often upregulated in cancer cells. In contrast, the tumour suppressor gene *CDH1* that codes for epithelial cadherin (E-cadherin) a cell adhesion protein that maintains epithelial integrity in differentiated tissues is lost in cancer.

It is becoming increasingly apparent that interactions between cancer cells and neoplastic stroma are involved in the complex processes of invasion and metastasis.

Cancer cells do not act autonomously but elicit the 'help' of associated stromal cells. One of these, the tumour associated macrophages (TAMs) are employed in cross talk with tumour cells, reciprocally stimulating each other. TAMs provide amongst other things the all-important matrix-degrading enzymes that facilitate tumour cell invasion. Thus, cancer progression cannot be seen purely in terms of oncogene gain and loss of tumour suppressor genes acting in isolation in the malignant cells. Stromal and infiltrating immune cells act in concert with the tumour cells to facilitate invasion and metastasis. Invasive and metastatic tumour cells not only acquire loss of attachments but also develop alterations in their shape, something that is reflected in the grading of a tumour by pathologists. This process termed epithelial to mesenchymal transition or EMT is thought to be a key process in metastatic spread of tumour cells. Epithelial cancer cells lose adhesion to their neighbouring cells and to any basement membrane. They also lose their normal top to bottom polarity and cytokeratin expression and take on mesenchymal capabilities such as increased cell motility. These cells stop dividing whilst they invade and migrate by inducing a temporary senescence. Invasive cells with metastatic potential must survive the rigours of circulation before entrapment in a capillary bed and then extravasation into the distant organ. Disseminated cells that survive the rigours of circulation and settle down in the secondary organ to form a metastasis, can reverse the EMT and re-acquire both epithelial and proliferative functions. This is mesenchymal to epithelial transition (MET). Figure 6.1 shows how a malignant cell takes on mesenchymal properties to escape from its place in an epithelium and invade adjacent and then distant tissues. Look again at Figure 2.3 to see the effects of MET on colorectal cancer in a lymph node. Metastatic deposits in a secondary site often show similar histology to the primary, enabling the pathologist to identify the site of the original tumour.

❯ See more about EMT in Chapter 6.

Genome instability, mutation, and epigenetic changes

Genomic instability is a key hallmark of cancer and single cell genomic analysis is revealing tumour evolution pathways and the key mutations that fuel invasion and metastatic spread. Generation of a mutator phenotype coupled with Darwinian selection of the fittest appears to aid the acquisition of several hallmarks that cancer cells acquire. Genomic instability can be achieved by several means including inherited mutations in genes required for maintenance of genomic integrity such as *BRCA1/2*, increased susceptibility to DNA mutation by environmental mutagens, sporadic somatic mutation, or epigenetic changes that alter tumour suppressor gene or oncogene expression. Loss of caretaker functions by genes involved in surveillance of DNA integrity can enhance the rate of subsequent mutations, fuelling tumour progression. Of particular importance is the loss of *TP53* that detects DNA damage and induces repair gene expression or apoptosis. Another source of genomic instability is afforded by loss of telomeric DNA and associated chromosomal rearrangements, amplifications, or deletions.

It is not just the loss of key multifunctional tumour suppressor genes that is important in promoting genome instability, but also the acquisition of certain oncogenes that drive rapid proliferation, such as *MYC* and *RAS*. Such oncogenes orchestrate multiple hallmark capabilities including angiogenesis, invasion, and replicative immortality. Distinct mechanisms are acquired at different times and by different tumours, but all require induction and tolerance of genomic instability to provide the mutations available for selection.

❯ See more about telomeres in Chapter 4.

Replicative immortality

The DNA replication machinery fails to duplicate completely the DNA at ends of chromosomes (telomeres). Thus, telomeres, composed of tandem hexanucleotide repeats, shorten at each cell cycle. This is known as the 'end replication problem'. Primary cells in culture go into a state of 'crisis' and die when their telomeres become critically short. This means that normal cells have a finite number of times they can replicate their genomes before they senesce. This replicative senescence in itself is a strong tumour suppressor mechanism.

Recognition of telomere erosion and the induction of senescence are carried out by two major tumour suppressor genes, *TP53* and *CDKN2A* (the gene locus coding for p16^{ink4a} and p14^{arf}). Loss of function of these is seen in most cancers. Thus, delayed p53-mediated surveillance or early loss of p16^{ink4a} and p14^{arf} due to deletion of the *CDKN2A* gene may allow cells with short telomeres to continue replicating, thus fuelling tumour progression. Cells with critically short telomeres form unstable and potentially mutagenic chromosome fusions.

Cancer cells require a large replicative potential to be cultured indefinitely or to generate a clinically significant tumour and must therefore acquire a means of replacing lost telomeric sequence. This is usually provided by the enzyme telomerase that tops up the sequences at the ends of chromosomes thus counteracting the progressive telomere erosion. Telomerase not only functions to maintain telomere DNA but also can have non-canonical functions, enhancing proliferation and resistance to apoptosis.

An alternative lengthening of telomeres method (ALT) occurs in some tumours, usually those of mesenchymal origin (sarcomas). This mechanism employs a DNA recombination strategy resulting in telomere length heterogeneity including some characteristically very long telomeres. ALT tumours are detected by staining for characteristic nuclear bodies called APBs (ALT associated PML bodies). These are nuclear entities comprising telomeric DNA and the promyelocytic protein (PML).

However, although telomere maintenance occurs in most cancer cells providing them with potential immortality, unchecked telomere erosion does occur in tumours lacking a telomere maintenance mechanism (TMM). Ever shorter telomeres are a potential source of genomic instability and mutation and confer a poor prognosis. To illustrate this, in spite of intensive treatment, patients with neuroblastomas that lack a TMM can still have a poor survival rate.

❯ See more about telomere maintenance in Chapter 4.

Reprogramming energy metabolism

High levels of proliferation require not only high levels of energy from fuel but also the means to generate cellular intermediates as building blocks for DNA and protein. Normal, non-dividing cells use the energy efficient mitochondrial oxidative phosphorylation to process glucose. Cancer cells sacrifice this 18-fold greater efficiency for the provision of anabolic intermediates by undergoing aerobic glycolysis. This is known as the Warburg effect and necessitates an increase in glucose or other fuel uptake, which is achieved in part by upregulation of cell membrane transporters such as the major glucose transporter, GLUT1 (SLC2A1). Positron emission tomography (PET) scanning can measure the level of glucose uptake by a tumour *in vivo* as shown in Figure 2.4. A positron emitting labelled glucose analogue is given to the patient and the gamma rays emitted from this tracer are used to image uptake and metabolism of glucose by a tumour. Gain of function mutations in isocitrate dehydrogenase 1 (*IDH1*) in gliomas and other tumours alter energy metabolism and these can be imaged in tissue sections using antibodies specific for the mutant form of the enzyme.

❯ See more about metabolic pathways in Chapter 4 and 5.

Evasion of the immune system

The immune system should recognize tumour cells as foreign and eliminate them. The failure of this immune surveillance with regard to tumour cells implies mechanisms of evasion are a hallmark of cancer cells. Immune suppression is associated with an increased likelihood of developing certain cancers. This was particularly relevant in untreated HIV positive patients who frequently developed the rare cancer of Kaposi sarcoma. Some transplant recipients have been found to succumb to donor-derived cancers, suggesting that in healthy donors, the pre-malignant cells were held in check but in the immune-suppressed recipient, these cells are free to develop into full-blown cancer.

The tumour microenvironment and tumour-promoting inflammation

Tumours are organs composed of many different tissues such as blood vessels, immune cells, and stroma to name a few. The importance of the tumour microenvironment in growth and progression is being increasingly recognized. The historical view that a tumour comprises only a relatively homogenous collection of tumour cells is being overturned. We now know that hyper-proliferation, genomic instability, and selection of the fittest tumour cells all fuel intra-tumour heterogeneity. Histopathologists have long recognized this heterogeneity of malignant cell morphology. Recent reports have also highlighted the ability of some tumour cells to trans-differentiate into endothelial like cells that form a malignant neo-vasculature or stromal cells for support without relying on recruitment of normal counterparts.

Infiltrating immune cells are widespread components of tumour microenvironment and usually associate with poorer prognosis. In fact, tumours have been described 'as wounds that never heal' because of the persistent presence of inflammatory cells in tumours. Many of these inflammatory cells have tumour-promoting activities such as supporting neo-angiogenesis, tumour cell

Figure 2.9 The tumour microenvironment and the cancer stem cell niche. Cancer cells in a niche comprised of vasculature and stromal support cells along with a wide range of infiltrating immune cells. Mesenchymal stem cells (MSC), endothelial progenitor cells (EPCs), and various bone marrow derived cells (BMDC).

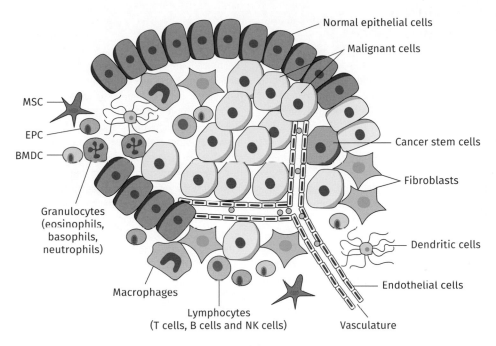

Figure 1 Upreti, M., et al. (2013). Trends in Cancer Research, 2(4): 309. http://tcr.amegroups.com/article/viewFile/1549/html/8801.

survival, and invasion. Genetic changes in cancer cells result in disruption of the normal intercellular communications, architecture, and function of neighbouring and distant non-neoplastic cells. This facilitates a microenvironment or niche for the growth and sustainability of a tumour.

Figure 2.9 shows the possible mixture of cells in a cancer niche. Blood vessels comprised of a lining of endothelial cells surrounded by smooth muscle cells, stromal fibroblasts, and certain immune cells support tumour cell growth and survival. Cancer stem cells have the potential to replace cancer cells.

❯ See more about cancer stem cells in Chapters 1 and 6.

 Key Points

- Hallmarks of cancer describe the essential biological processes necessary for tumour formation.
- The hallmarks must be taken in context of a microenvironment of malignant cells and recruited non-malignant cells, as both contribute to tumour development and progression.

Bigger picture 2.1
A holistic view of cancer

The only way to improve treatment and patient management is to increase our understanding of cancer and to do this it is important to understand the hallmarks of cancer. It is imperative when designing or analysing cancer studies to compare like with like. This entails some knowledge of cancer pathology and nomenclature. The hallmarks of cancer can only be fully appreciated in the context of tumour pathology.

Many of the previous research and diagnostic approaches to cancer have been oversimplified by treating cancer as a disease of just tumour cells. This approach fails to evaluate the heterogeneity of cancer cells and their interactions with the microenvironment as a whole. Tumour cells interact with other tumour cells and host cells in a microenvironment that adapts to ensure survival and growth of the cancer. Other cells present will be infiltrating immune cells and support cells such as stroma and vasculature. Traditionally measurements are made on the mass of cells in a tumour, only some of which will be cancer cells. Many studies take surgical material from tumours, homogenize it, and measure genetic changes or molecular expressions in the resultant soup. This approach not only fails to recognize the varied contribution made by non-neoplastic cells but also dilutes the information on specific changes occurring in the tumour cells. *In-situ* techniques, such as RNA-hybridization and immunohistochemistry (IHC) enable specific and subtle changes in expression to be measured. Understanding the interrelationships between cell types should lead to novel and targeted approaches to treatment. New generation technologies are making single cell analysis and whole genome sequencing more affordable. Liquid biopsy and measuring cancer specific genome changes is facilitating the tracking of cancer progression.

? Pause for thought

1.   What are the advantages and disadvantages of studying clinical samples and data compared to *in vitro* studies on cells, for increasing our understanding of cancer?

Chapter Summary

-   A basic knowledge of cancer pathology and its terminology including staging and grading of tumours is essential to the understanding of cancer.
-   Tumours can be benign or malignant. Malignant tumours are invasive and are called cancer.
-   A tumour is composed of many types of cell including non-malignant as well as neoplastic cells.
-   Cancer cells not only proliferate and escape normal growth and survival restrictions but also recruit an environment or niche that supports their requirements.
-   A series of fundamental principles of cancer biology are called the hallmarks of cancer. Without an appreciation of cancer biology, we will be unable to improve diagnostics, prognostics, and therapy.
-   The principles of cancer progression are applied to population screening for certain cancers. Cervical cancer and colorectal cancer provide examples that illustrate these principles.

Further Reading

Fouad, Y. A. and Aaner, C. (2017). 'Revisiting the hallmarks of cancer'. American Journal of Cancer Research 7(5): 1016–36.

The authors revisit Hanahan and Weinberg's seminal paper and present their condensed and updated view of the Hallmarks of neoplasia. They point out that some of the original traits can be applied to benign neoplasms and therefore cannot be strictly considered hallmarks of cancer as the latter is a term for invasive malignancy

Hanahan, D. and Weinberg, R. A. (2011). 'Hallmarks of cancer: the next generation'. Cell 144: 646–74.

Hanahan and Weinberg's inspiring paper on the basic principles that govern cancers. These traits or 'hallmarks' are their revised list of fundamental features that link malignancies, based on data from *in vitro* and pathological study.

Krebs, M. G., Hou, J-M., Ward, T. H., Blackhall, F. H., and Dive, C. (2010). 'Circulating tumour cells: their utility in cancer management and predicting outcomes'. Therapeutic Advances in Medical Oncology 2(6): 351–65.

Diagnostic and prognostic value of circulating tumour cell analyses in patient management. The methodology and potential applications of the technique are explained in detail.

Malanchi, I. (2013). 'Tumour cells coerce host tissue to cancer spread'. BoneKEy reports 2 Article number 371. Doi: 10.1038/bonekey.2013.105.

Malanchi presents a very thorough account of how tumour cells hijack the functions of non-malignant adjacent cells to provide a suitable niche for their survival and malignant progression.

National Cancer Institute: http://www.cancer.gov

The US government principle site for cancer research. It gives accurate and up to date information on different kinds of cancer, their treatment, and involvement in clinical trials.

American Cancer Society: http://www.cancer.org/index

This is the website of the American Cancer Society.

Dictionary of Cancer terms: http://www.cancer.gov/dictionary

U.S. National Cancer Institute dictionary of cancer terminology.

 Discussion Questions

2.1 Describe what a pathologist requires for a report and how they achieve this.

Hint: think about grading and staging of cancers. Look at the possibilities of using immunohistochemistry to identify the probable cell of origin of a tumour.

2.2 What are the hallmarks of cancer, and how do they relate to benign and malignant tumours?

Hint: think about the differences between benign and malignant tumours. Read Fouad et al.'s 2017 article and the papers referenced therein.

2.3 What are the advantages and disadvantages of population screening for cancer?

Hint: think about the effects of possible false positive or false negative results.

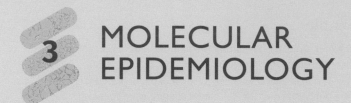

3 MOLECULAR EPIDEMIOLOGY

Learning Objectives

- describe global cancer trends including incidence, mortality, geographical variations, and gender variations;

- describe the different types of epidemiological studies used in evaluating associations between particular factors and cancer risk and the challenges inherent within these;

- explain the role of genetic and non-genetic factors in disease contribution and how these may interact to influence cancer risk; and

- explain how epidemiological and laboratory-based studies can be used to inform cancer development in specific cancer types including breast, lung, melanoma, and colorectal cancer.

This chapter considers the influence of genetics, environmental factors, and lifestyle behaviours on the risk of cancer developing. A risk factor is anything that increases the likelihood (probability) of an individual developing a particular disease. In cancer, risk factors can be broadly divided into two categories: intrinsic (genes inherited through the germ-line) and extrinsic factors. The latter include environmental factors (exposure to high-levels of radiation, cancer causing chemicals, and certain infectious agents) as well as lifestyle factors (such as smoking, an unhealthy diet, and unprotected exposure to sunlight radiation). Our understanding of this complex area is informed by epidemiological data supported by laboratory-based studies. In this chapter, we will review some of these data using selected examples of cancers to shed light on the relationship between extrinsic and intrinsic factors and how these influence cancer development.

3.1 Cancer trends

Cancer is a leading cause of disease and of death worldwide. Incidence and mortality data help us understand the burden of cancer and provides clues to the possible environmental, lifestyle, and other risk factors that contribute to cancer aetiology. This in turn, informs interventions that could be introduced to reduce cancer risk, for example, by reducing cigarette consumption, and earlier identification of disease through implementation of screening programmes. By comparing pre-and post-intervention incidence and mortality data, the effectiveness of any strategies put in place can also be measured.

Cancer incidence refers to the number of new cases diagnosed for a given period, usually a year. The data are expressed as a rate (number of cases per 100,000 population) and are commonly age-standardized. Age-standardized rates (ASR) take into account variations in the age structure of populations, and so allows comparisons between different geographical regions and different time-periods to be made. Similarly, mortality refers to the number of deaths occurring in a given period, in a specified population, and is age-standardized.

Cancer incidence and mortality statistics can be obtained from a variety of sources:

- The Global Cancer Observatory website provides information on the occurrence of cancer worldwide. It is held by the International Agency for Research on Cancer (IARC), a specialized agency of the World Health Organization. One of these databases, GLOBOCAN provides global incidence and mortality estimates for all major cancer types. These are calculated every four to six years.

- The National Cancer Institute (NCI) provides access to reports containing cancer incidence and mortality information across the USA. These are provided annually.

- Cancer Research UK (CRUK) provides access to cancer data for the UK and is updated typically every two years.

Data on cancer incidence and mortality are collated by cancer registries. Countries such as the UK and the USA have well-established systems for gathering this information but many countries, particularly low and middle-income countries do not have systems that register these data. As such, data quality varies greatly by region and this should be considered when interpreting incidence and mortality statistics.

Risk factor associations and cancer

To identify risk factors associated with cancer, types of epidemiological studies typically used are observational studies and randomized controlled trials. Observational studies are so called because the investigator observes individuals without manipulation or intervention. Such studies include case control studies and prospective cohort studies. This is in contrast to randomized controlled trials in which the investigator does intervene.

- In case control studies, individuals with a specific cancer (cases) recall previous exposure to a particular factor that is suspected of causing the disease (e.g. dietary habits) before they were diagnosed. These accounts are compared with accounts from healthy cancer free individuals (controls). Such studies can reveal differences that might suggest an association of the factor with the cancer.

- In prospective cohort studies, healthy individuals who display certain behaviours (e.g. smoking) are followed longitudinally over time to identify those that may eventually develop a cancer. Associations between the behaviour and the disease are then checked. This method reduces recall bias inherent in case control studies but requires large sample sizes and follow up durations.

- In randomized controlled trials, individuals are randomly assigned to one of two groups. One group receives a particular intervention (e.g. a nutritional supplement or drug) whilst the second (control) group does not receive the intervention. After a period of time, the results of the two groups are compared to establish the effect of the intervention. This is the simplest form of a RCT; it is possible to have multiple arms to a trial with groups randomized to various interventions. RCTs overcome selection bias inherent in the above two study types since participants are randomized to either the intervention or control arm of the experiment. Randomization can also control for confounding variables. However, RCTs can be limited by subject compliance and in the case of dietary interventions, consumption of atypical amounts of dietary supplements.

When interpreting epidemiological data it is important to understand the distinction between association and causation. Observational studies can only strictly study the association between an exposure (e.g. obesity) and outcome (e.g. bladder cancer). Associations identified from these studies can then be used to formulate hypotheses to be tested in subsequent controlled experiments which may be of various types: RCT, *in vitro*, or *in vivo*. Epidemiological studies when considered alongside biological studies help determine causal relationships. For more extensive descriptions of epidemiological principles, you are directed to read Bonita et al. (2006), listed in the references.

Global cancer incidence and mortality

The most recent published data available on global cancer trends is for 2018. These sources estimate that 18.1 million new cases of cancer and 9.6 million deaths will occur worldwide during that year. The most common are lung, female breast, colorectal, and prostate cancers, accounting for approximately 40 per cent of all cancers diagnosed. Around half of all cancer deaths annually are due to lung, liver, stomach, colorectal, and female breast cancers.

Worldwide cancer rates, however, mask variations in cancer burden in individual countries. For example, the highest colorectal cancer incidence rates occur in Australia/New Zealand, Europe, and North America, whilst the lowest rates are found in Africa, some Asian countries, and Latin America. Similarly, female breast cancer incidence rates are highest in Australia/New Zealand, Europe, and North America, and lowest in Africa and Asia.

Cancer incidence rates are generally two- to three-fold higher in high-income countries (HICs) than in low- and middle-income countries (LMICs). Mortality trends are reversed with cancer rates higher in LMICs compared to HICs. The number of cancer cases is predicted to rise as populations grow and age. Globally, almost 29.5 million new cancer diagnoses are predicted by 2040, a 63 per cent increase on the 18.1 million new cases estimated in 2018. The greatest proportional increases are predicted to be in LMICs. A broad explanation is that LMICs, as they undergo economic transition, will adopt lifestyle behaviours that increase cancer risk. These include smoking, excess body weight, physical inactivity, and altered child-bearing patterns. As such the pattern of cancer incidence is likely to follow that seen in HICs with a decline in infection or poverty related cancers (e.g. cervix, stomach, liver) and an increased burden of cancers associated with smoking, reproductive, and dietary risk factors (lung, breast, and colorectal).

Age and lifetime cancer risk

Cancer is primarily a disease of older people, with incidence rates increasing with age for most cancers. Figure 3.1 provides estimated incidence rates for all cancers combined (excluding non-melanoma skin cancer) by age group for men and women. Only in the youngest age group (up to fourteen years) are rates similar between the genders. After childhood, rates are higher in women than in men until about the age of fifty years. This trend is reversed from the age of sixty years when incidence rates in men overtake those in women. The higher rates in women before the age of fifty years can be explained by the relatively earlier age at onset of cervical cancer and breast cancer compared with other major cancers. Above the age of sixty years, prostate cancer and lung cancer in men become more common.

Figure 3.1 Age and cancer risk. Estimated incidence rates for all cancers combined (excluding non-melanoma skin cancer) by age group for men and women. See text for further details.

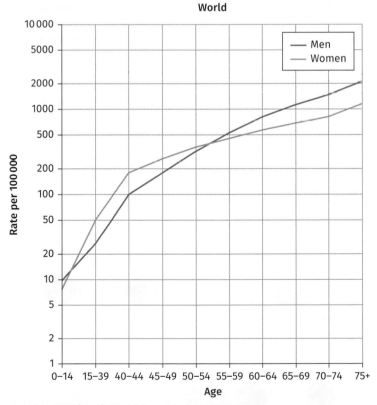

Figure I.1.8 in IACR (2014). World Cancer Report 2014.

The lifetime risk of developing cancer is the probability that a person will be diagnosed with cancer over the course of his or her lifetime. It is commonly expressed as a percentage (e.g. 20, 33, or 50 per cent) or using odds (e.g. one in five, one in three, or one in two, respectively). For example, a woman with no known risk factors for breast cancer has a lifetime risk of about 12 per cent of developing breast cancer. This equates to one out of every eight women developing breast cancer during their lifetime. In the UK, the lifetime risk of developing cancer has increased from one in three for those born in 1930 to one in two for those born after 1960. Much of this increase is due to people living longer, but an increase in the incidence of particular cancers has also had an impact. Of note are increases in breast cancer amongst women due to lifestyle changes such as women having fewer children and at later ages, as well as increased detection through breast cancer screening programmes. Similarly in men, implementation of screening programmes and dietary changes (increased red meat consumption) have increased the incidence rates of prostate and colon cancer, respectively.

3.2 Inherited cancers

As described in Chapter 1, cancers can be familial (inherited) or sporadic. In familial cancers, the disease causing mutation is carried in the germline and so is present in all cells of the body at birth. This does not cause cancer on its own and additional somatic mutations are still required for cancer to develop. However, the risk of cancer developing is much higher in individuals that carry a germline mutation compared to the general population.

> See Chapter 1 for introduction to familial cancers.

For most cancers, the **relative risk** for that cancer is two to four times higher if a first-degree relative (parent, sibling, or child) has been affected. For example, a woman with a first-degree female relative with breast cancer has a two-fold higher risk of developing breast cancer compared to a woman without a family history. Disease onset is likely to be earlier, and such individuals are also at risk of developing multiple primary tumours. Evidence for the heritability of cancer comes from studies of families covering two or more generations in which a particular tumour or set of tumours aggregate. Figure 3.2 shows an example pedigree of inherited breast cancer.

Figure 3.2 Example of an inherited breast cancer pedigree. Solid circles, females with breast cancer; open circles, females without breast cancer; open squares, males without breast cancer. Diagonal line through symbols represent diseased individuals.

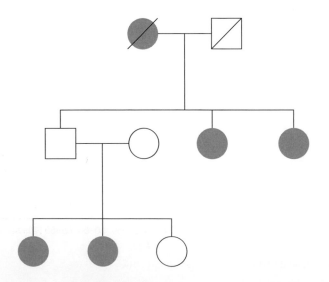

Comparison of the concordance of cancer between monozygotic and dizygotic twins also supports a genetic component to cancer risk. A significantly higher concordance of cancer at the same anatomical site is observed for monozygotic twins compared with dizygotic twins. Many genes have now been identified that contribute germline susceptibility to cancer. Table 3.1 summarizes some

Table 3.1 Examples of familial (hereditary) cancer syndromes. Only exemplar germ-line mutations are included in this table. Many of these syndromes show genetic heterogeneity and so alternative genes may be associated with the syndrome. Refer to OMIM for fuller information about these syndromes and associated germ-line mutations. See also Chapter 1.

Cancer syndrome	Germ-line mutation	Inheritance mode	Normal gene function	Common cancer type(s)
Ataxia telangiectasia	ATM	Autosomal recessive	Repair of damaged DNA	Childhood lymphomas and leukaemia and breast cancer
Familial adenomatous polyposis	APC	Autosomal dominant	Negative regulator of the pro-oncogenic WNT/β-catenin signalling pathway	Colorectal cancer
Fanconi anaemia	FANCA	Autosomal recessive	Repair of damaged DNA	Breast cancer
Hereditary breast and ovarian cancer	BRCA1 BRCA2	Autosomal dominant	Repair of damaged DNA	Breast and ovarian cancers
Hereditary retinoblastoma	RB1	Autosomal dominant	Cell cycle regulator	Retinoblastoma
Li-Fraumeni syndrome	TP53	Autosomal dominant	p53 transcription factor, multiple roles including induction of cell cycle arrest and apoptosis	Breast cancer, osteosarcoma and soft tissue sarcomas (e.g. muscle), brain cancers, and leukaemia
Lynch syndrome	MSH2 MLH1	Autosomal dominant	Involved in DNA mismatch repair	Hereditary nonpolyposis colorectal cancer
Multiple endocrine neoplasia type II	RET	Autosomal dominant	Receptor tyrosine kinase involved in MAPK and PI3K signalling pathways	Endocrine tumours: thyroid, adrenal, and pancreas
Peutz-Jeghers syndrome	STK11	Autosomal dominant	Regulates cell metabolism and energy homeostasis as well as cellular stress response to DNA damage	Gastrointestinal tumours, pancreas, cervix, ovary, and breast cancers
von Hippel-Lindau syndrome	VHL	Autosomal dominant	E3 ligase involved in protein degradation, e.g. Hif under normal oxygen conditions	Haemangioblastomas, (blood vessel tumours of the brain, spinal cord, and eye), renal cell carcinoma, pancreatic cancer
Xeroderma pigmentosum	XPA XPB XPC	Autosomal recessive	DNA excision repair proteins	Skin cancers, leukaemia, brain, and lung tumours

of these mutations and associated cancer syndromes. Most hereditary cancer syndromes follow an autosomal dominant inheritance pattern, in which inheritance of one mutant allele of a cancer-causing gene is sufficient to predispose that individual to the cancer syndrome. A small number of cancer syndromes follow an autosomal recessive inheritance pattern in which inheritance of two mutated copies of a cancer-causing gene is required to predispose the individual to developing the cancer syndrome. Although two mutations predispose to the syndrome, inheritance of one mutated allele can still increase cancer risk. For example, inheritance of two mutated alleles of *ATM* predisposes the individual to ataxia telangiectasia, whilst inheriting one mutated *ATM* allele predisposes to breast cancer and possibly others.

Individuals who carry germline mutations associated with hereditary cancer syndromes face a high probability of developing the cancer but not always a 100 per cent certainty. For example, individuals who carry mutations in the *BRCA1* gene have approximately 70–85 per cent chance of developing breast cancer. Similarly carriers of a mutated *APC* gene are at a 95 per cent risk of developing familial adenomatous polyposis, leading to colorectal cancer.

High penetrance genes or alleles such as *BRCA1* and *APC* are rare; and explains cancer risk only in a proportion of families in which the disease clusters. The remaining cancer risk is unclear but it is becoming increasingly likely that a substantial proportion of hereditary cases are polygenic caused by a combination of moderate or low risk genetic variants. Individually these genetic variants confer only a small risk but several moderate or low risk variants working together and in conjunction with environmental and lifestyle influences can increase cancer risk substantially. Moderate and low penetrance alleles tend to occur at higher frequency within the general population compared to high penetrance genes. The relationship between frequency of the allele and relative risk is presented graphically in Figure SA 3.1 for breast cancer.

Single nucleotide polymorphisms (SNPs) are thought to account for the majority of moderate and low penetrance cancer susceptibility genetic variants. A SNP is characterized by alleles that differ at a single nucleotide between two individuals or between paired chromosomes in the same individual. To be classified as a SNP, the variation must occur in at least 1 per cent of the population. Over a 100 million SNPs are estimated to occur distributed throughout the human genome in both protein coding and non-coding regions. Their functional consequence depends on the nature and location of the sequence change. Some have no effect whilst others may alter protein function, the protein levels, or produce non-functional proteins. Genome-wide association studies (GWAS) can be used to identify SNPs that may predispose to cancer. This is achieved by comparing the genome sequences of a large group of people affected by a particular cancer with a second large unaffected group (the control). If certain variants are found to occur more frequently in the cancer cases than in the control cases, then these variants are said to be associated with the disease. In GWAS, SNP genotyping technologies are used to genotype hundreds of thousands of SNPs simultaneously. Polymorphisms that show statistically significant levels of association are subjected to further rounds of testing in cancer-cases and control groups. Scientific Approach Panel 3.1 discusses some of the concepts described in this section using hereditary breast cancer as an example.

Scientific approach panel 3.1
Discovering hereditary breast cancer susceptibility genes

Hereditary breast cancer accounts for approximately 10 per cent of all breast cancer cases. Of these, up to 35 per cent of cases are due to one of the few mutated, rare, but highly penetrant genes including *BRCA1*, *BRCA2*, *PTEN*, and *TP53*. An additional 2–3 per cent of hereditary cases are due to mutations in moderate penetrance genes (e.g. *CHEK2*, *ATM*, and *PALB2*) and the rest are thought to involve a combination of low penetrance susceptibility genes each conferring a small increased risk individually but a greater risk in combination.

BRCA1 was the first major gene to be identified associated with hereditary breast cancer. This gene is inherited in an autosomal dominant manner and its prevalence varies by race and ethnicity; one in 400–500 individuals of Western European descent carry this gene but in the Ashkenazi Jewish population it is much higher with one in forty-five carriers. Onset of breast cancer in carriers is typically before the age of fifty, and the cancer is often bilateral, occurring in both breasts. Carriers are also at increased risk of developing other cancers such as ovarian and prostate cancer. Although inherited in an autosomal dominant manner, *BRCA1* acts recessively as a tumour suppressor and so the second wild-type allele is lost through somatic changes.

In 1990, Jeff Hall et al., using DNA samples from twenty-three extended families with 146 cases of breast cancer and 329 participating relatives, linked early onset breast cancer susceptibility to a region on the long arm of chromosome 17 (17q21). The method used was linkage analysis, which works on the basis that genetic markers that reside physically close to each other on the same chromosome are co-inherited; and so by analysing genetic markers at known locations, disease genes in close proximity can be identified. Further studies confirmed the precise location of the *BRCA1* gene as 17q21.31 and subsequently in 1994 a group of researchers isolated and sequenced *BRCA1* using positional cloning methods. Sequence analysis revealed that of the five families in which

Figure SA 3.1 Relative risk and frequency allele for breast cancer genes. Alleles occurring at low frequency tend to be associated with higher relative risk (e.g. *BRCA1*) and those occurring at higher frequency are associated with lower relative risk (e.g. *FGFR2*). High penetrant genes are associated with a RR of > 4, moderate penetrant genes with a RR of 2–4, and low penetrant genes with a RR <1.3.

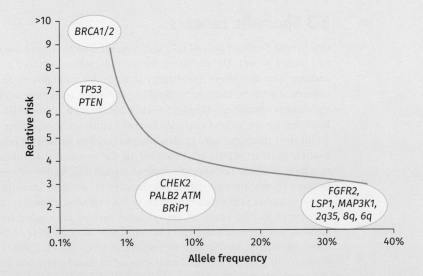

Figure 1 in Fanale et al. (2012). Oncogene, 31: 2121–8.

BRCA1 aggregated, each family carried a unique mutation in the gene: an eleven-base pair deletion, a one-base pair insertion, a stop codon, a missense substitution, or a mutation localized in the regulatory region of the gene. We now know that *BRCA1* is very complex, twenty-two exons in length and that a wide spectrum of mutations can arise, dispersed throughout the gene. These result in lower protein amounts or production of truncated molecules with impaired function. Wild-type BRCA1 is involved in repair by homologous recombination (HR), a pathway that is activated when double-stranded DNA breaks occur. DNA double-stranded breaks are detected by ATM, a kinase that phosphorylates a number of proteins including BRCA1. The exact role of BRCA1 in HR is only now being defined with Weixing Zhao and colleagues showing in 2017 that BRCA1 complexes with BARD1 also a tumour suppressor protein to repair DNA in the latter stages of the HR repair pathway. Non-functional BRCA1 leads to defects in the repair of damaged DNA, increasing genomic instability, and thus accelerating the rate at which the genome acquires mutations.

Linkage analysis cannot identify moderate or low penetrance alleles; these have been identified either through association studies or more recently through genome-wide sequencing or exome sequencing. A gene identified through association studies is *CHEK2*. In 2002, The CHEK Breast Cancer Consortium analysed the frequency at which the *CHEK2*1100delC* variant is carried in families with breast cancer. These authors used a total of 1,620 controls (cancer-free cases) and 1,071 breast cancer cases (drawn from 718 *BRCA1* or *BRCA2* negative families). *CHEK2*1100delC* was detected in 1.1 per cent of the control group but at much higher frequency (5.1 per cent) in individuals with breast cancer without mutated *BRCA1* or *BRCA2*. The authors estimated that breast cancer risk for carriers of this variant is increased by twofold. CHEK2 is a cell cycle checkpoint kinase. In response to DNA damage, CHEK2 is phosphorylated by ATM, which then goes on to phosphorylate p53 and BRCA1. These in turn promote cell cycle arrest and the activation of DNA repair pathways. The *CHEK2*1100delC* variant results in a non-functional truncated protein leading to impaired cell cycle arrest and DNA repair, thus contributing to tumourigenesis.

Whole-exome or genome sequencing is now increasingly being used as the method of choice for detecting cancer susceptibility genes. One gene identified through whole exome sequencing of familial breast cancer cases is *XRCC2*. *XRCC2* encodes a protein involved in repairing double-stranded DNA breaks, and is mutated in some familial breast cancer cases through truncating or point mutations, leading to loss of protein function.

3.3 Sporadic cancers

Only about 5–10 per cent of all cancers result directly from gene defects inherited from a parent. The rest occur through mutations acquired in somatic cells, and are sporadic. These mutations can arise due to exposure to certain environmental factors, lifestyle behaviours, and some infectious agents as well as spontaneous mutations arising, for example through errors in DNA replication. Evidence for the role of extrinsic factors contributing to cancer risk has been established through many epidemiological studies as well as laboratory-based studies, some of which are described in the sections that follow. The extent to which extrinsic factors contribute to cancer risk varies by cancer type. For instance, it is estimated that approximately 90 per cent (nine in ten cases) of non-melanoma skin cancers are due to UV radiation exposure, at least 75 per cent (three in four cases) of oesophageal cancers are caused by tobacco and alcohol, and approximately 90 per cent (nine in ten cases) of cervical cancer cases are caused by the human papilloma virus (HPV). Table 3.2 provides a summary of cancer types and examples of environmental and lifestyle factors

Table 3.2 Examples of cancers and associated extrinsic risk factors. For dietary components, see Figure 3.4

Cancer type	Examples of potential extrinsic risk factors
Breast	Diet, alcohol, obesity
Cervical	Human papilloma virus
Colorectal	Diet, smoking, alcohol, obesity
Head and neck	Tobacco, alcohol
Hepatocellular	Hepatitis B virus, hepatitis C virus
Lung	Smoking, air pollutants, occupational exposure to carcinogens (e.g. asbestos).
Oesophageal	Smoking, alcohol, obesity
Prostate	Diet, obesity
Skin	Sun exposure (UV radiation)
Stomach	Alcohol, obesity, diet, *Helicobacter pylori*

that contribute to cancer risk. Some of these risk factors—sunlight exposure, smoking, diet, and obesity—are described below.

Environmental exposure—UV light and tobacco smoke

Exposure to UV light plays a very significant role in the formation of skin cancers. Skin cancers can be categorized as melanomas or non-melanomas, dependent on the cell type in which they originate. Melanomas arise in melanocytes, which are responsible for making the pigment melanin, whilst non-melanoma skin cancers arise in the outermost layers of the skin, in basal or squamous cells. Studies have shown that exposure of mice to ultraviolet B (UVB) wavelengths (280–320 nm) can induce skin tumours in mice and the effect is much stronger than with exposure to UVA (320–400 nm). Case control studies which have gathered sun exposure information (e.g. time spent sunbathing, frequency of sunburns) also support an association between exposure to sun rays and skin cancers, particularly in fair-skinned populations. UVA and UVB radiation induces the formation of dimers between adjacent pyrimidines and promotes the production of oxygen and nitrogen reactive species which go on to cause damage to DNA, lipids, and proteins. If uncorrected by DNA repair mechanisms, mutations in key cancer-causing genes can arise, driving tumour formation.

A variety of cancers are associated with exposure to certain chemicals, including asbestos used in construction and car manufacturing, benzene used in the petrochemical industry, and chemicals found in tobacco smoke. Tobacco smoke is responsible for 85–90 per cent of all lung cancer cases and contains at least fifty chemicals that are carcinogenic. The most potent of these are polycyclic aromatic hydrocarbons (PAH), N-nitrosamines, and aromatic amines. The carcinogenic effect of tobacco smoke on the lungs was shown in the 1950s in mice. Exposure of mice to PAH and NKK (a specific N-nitrosamine) increased the number of lung tumours in mice compared to non-treated controls. An immense amount of epidemiological evidence accumulated from case control and cohort studies has shown an increased risk of lung cancer with tobacco, and also with others cancers including of the bladder, pancreas, oesophagus, larynx, mouth, and kidney. Smokers have at least twenty-fold increased risk of developing lung cancer compared to a life-long non-smoker. The risk is dose and time dependent with lung cancer risk increasing with the number of years of smoking and the number of cigarettes smoked per day. Exposure to smoke second-hand (passive smoking) also increases the risk of developing lung cancer in non-smokers.

Mutational signatures

Tobacco carcinogens exert their carcinogenic effects by binding to DNA and forming covalent binding products called DNA adducts. At these sites, the potential for mutations are increased leading predominantly to cytosine to adenine transversions (C>A) a sequence change that also dominates in lung cancer patients. This is an example of a mutational signature. Mutational signatures can be viewed as fingerprints left on cancer genomes by different mutagenic processes. Analysing these signatures allows us to identify which environmental, dietary or occupational exposure (risk factor) may have created the mutation and therefore contributed to the formation of the cancer. Initially mutational signatures were explored through a small number of frequently mutated genes, notably *TP53* in skin and lung cancers where a link between environmental exposure and cancer risk had been established. These early studies found two examples of mutagen specific mutation patters: the genome of melanoma skin cancers associated with exposure to UV light contained predominantly C to T transitions (substitutions) or CC to TT dinucleotide substitutions both occurring at dipyrimidines. This change is almost never found in human tumours not related to sunlight. Similarly, C to A transversions dominate in smoking-associated lung cancers, but are found much less frequently in lung cancers of non-smokers, or in other cancers not related to smoking.

As we know from Chapter 1, it has been possible to read the mutational records of whole genomes derived from cancer patients using next generation sequencing technologies. Using pooled data sourced from catalogues of somatic mutations, in 2013, Alexandrov and colleagues conducted a large-scale study analysing 4,938,362 mutations from >7,000 patients across thirty different cancer types. These authors applied an algorithm to extract mutational signatures based on a ninety-six-mutation classification system. This is calculated based on the possibility of six substitution changes occurring: C>A, C>G, C>T, T>A, T>C, and T>G (all substitutions here are referred to by the pyrimidine of the mutated Watson–Crick base pair) together with changes in the bases immediately 5′ and 3′ to the mutated nucleotide, yielding the possibility of ninety-six different trinucleotide sequences. Their work revealed twenty-one distinct mutational signatures, some of which were present in many cancer types and others confined to a single cancer class. Currently there are thirty known and validated mutational signatures: designated signatures 1–30, mapped across forty cancer types. The patterns of these mutational signatures and up to date information about them can be found on the COSMIC website. Figure 3.3 provides information on a subset of these mutational patterns including signatures 4 and 7, which are attributed to tobacco smoke and exposure to UV light, respectively.

Within individual cancers, there can be multiple mutational processes operating, each contributing its own mutational pattern. Thus the final catalogue of mutations will contain a mixed record of various past exposures, both intrinsic and extrinsic. By applying mathematical models and computational tools, it is possible to extract meaningful signatures so that the contributions made by various mutagenic processes within individual cancers can be identified.

❯ See Chapter 1 for cancer genome analysis.

Figure 3.3 Mutational signatures in human cancers and proposed cause. (a) The presence of mutational signatures across thirty cancer types (only seven of the thirty identified signatures are shown. For full list, see the COSMIC website). Cancer types are ordered alphabetically in columns and signatures displayed in rows. Proposed cause listed adjacent to the signature row. Adapted from Figure 3 in Alexandrov et al. (2013), Nature 500(7463): 415–21 and the Cancer Genome COSMIC site (https://cancer.sanger.ac.uk/cosmic/signatures). (b) Signature 4 is associated with exposure to tobacco smoke and signature 7 is associated with exposure to UV. The pattern of each signature is displayed according to the ninety-six substitution classification (i.e. examination of six substitution types: C>A, C>G, C>T, T>A, T>C, and T>G, as well as examination of the bases immediately 5′ and 3′ to each mutated base generating ninety-six possible mutation types (six types of substitution * four types of 5′ base * four types of 3′ base)). The mutation types are on the horizontal axes (six substitution types displayed in different colours), whereas vertical axes depict the percentage of mutations attributed to a specific mutation type. All mutational signatures are displayed based on the trinucleotide frequency of the reference human genome version GRCh37.

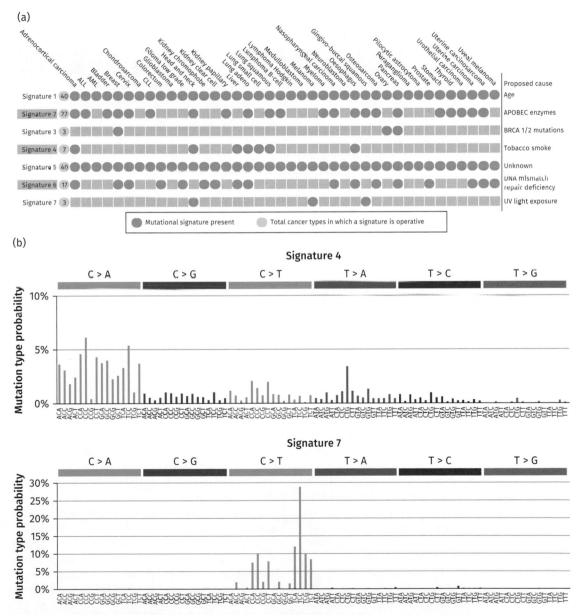

Adapted from Figure 3 in Petljak and Alexandrov (2016). Carcinogenesis, 37(6): 531–40.

Gene–environment interactions

The impact of environmental factors on an individual's risk of developing cancer may be modified by their genetic background. The converse is also true: an individual's inherited susceptibility to cancer could be modified by lifestyle or environmental exposure. This type of effect modification is described as gene-environment interactions. Staying with our example of lung cancer, a high proportion of smokers, but not all, develop lung cancer. Inherited genetic polymorphisms in specific genes may account for some of this difference. The association between genetic polymorphisms, environmental factors, and cancer risk in the context of lung cancer is described in Scientific Approach Panel 3.2. Table 3.3 lists some of the genetic polymorphisms that have been associated with specific environmental factors to increase or decrease cancer risk. A number of these—*CHRNA5* and *CHRNA3*, *ADH* and *ALDH*, and *MTHFR* variants—are discussed in the sections that follow.

 Key Points

- A small proportion of cancers are due to germline inheritance of mutated genes. These cancers tend to be associated with earlier age onset, aggregate in families, and affected individuals can present with more than one cancer type.
- The vast majority of cancers are sporadic, with mutations occurring in somatic cells caused by exposure to carcinogens through the environment, due to lifestyle behaviours or mutations occurring randomly through, for example, errors in DNA replication.

Scientific approach panel 3.2
Genetic polymorphisms, smoking, and lung cancer risk

In 2008, three independent GWAS for lung cancer were published: one conducted by the International Agency for Research on Cancer (IARC) involving 1,989 lung cancer cases and 2,625 controls spanning six Central European countries, a second by MD Anderson (Houston, TX, USA) including 1,154 cases and 1,137 controls, and a third from DeCode, Iceland, including 665 cases and more than 10,000 controls. All three studies identified a locus in chromosome region 15q25 to be strongly associated with lung cancer. This susceptibility region contains genes for the nicotinic acetyl choline receptor (nACHR) subunits 3 and 5 (*CHRNA3* and *CHRNA5*). Such subunits are expressed in neurons and in cells of the lungs. The IACR and the MD Anderson studies identified a direct association with three SNPs in this region and lung cancer. One of these encodes an amino acid substitution from aspartic acid to asparagine at residue 398 (D398N) in the *CHRNA5* gene. The Icelandic study identified an association with the same genetic region but with smoking quantity, concluding that the variant increases lung cancer risk indirectly through smoking behaviours (e.g. duration and dose of lifetime smoking). nACHRs are transmembrane receptors that are activated by the neurotransmitter acetylcholine but also by nicotine and its metabolites. As nACHRs are believed to play a role in nicotine dependence, it is possible that variant receptors might increase addiction to tobacco and thereby increased exposure to tobacco carcinogens.

The 2008 studies were the first GWAS to attempt to find a genetic component for a disease that has a strong environmental cause. Since then, further large-scale association studies have identified additional lung cancer susceptibility loci. Many SNPs have now been located across multiple cancer types including breast, prostate, colorectal cancer, and others. You can search for these at the GWAS catalogue maintained by the National Human Genome Research Institute (NHGRI) and the European Bioinformatics Institute (EBI).

Table 3.3 Examples of genes (polymorphisms) that interact with environmental factors to modify cancer risk

Gene	Environmental exposure	Risk of cancer type(s)	Variant activity
ADH1B Alcohol dehydrogenase	Alcohol	Oesophageal cancer	Polymorphisms encoding ADH enzyme with high activity associated with increased risk (see text)
ALDH2 Aldehyde dehydrogenase	Alcohol	Oral, pharyngeal, laryngeal, and oesophageal cancers	Polymorphisms resulting in the production of enzyme with reduced activity associated with increased risk (see text)
CHRNA5/A3 Cholinergic receptor nicotinic α5 subunit/α3 subunit	Cigarette smoking	Lung cancer	Specific genetic variants associated with increased risk of lung cancer amongst smokers (see text)
COX2 Cyclooxygenase 2	*Helicobacter pylori*, cigarette smoking	Gastric cancer	Specific 765G>C polymorphism with exposure to *H. pylori* infection or cigarette smoking increases gastric cancer risk
CYP1A1 Cytochrome P450 1A1	Cigarette smoking	Lung cancer	Specific variants at *CYP1A1/A2* genes associated with increased risk of lung cancer in smokers
LSP1 Lymphocyte specific protein 1	Number of births	Breast cancer	*LSP1* variants can modify the risk of breast cancer by number of births
MTHFR Methylenetetra hydrofolate reductase	Dietary folate intake	Breast cancer Endometrial cancer	677C>T variant is protective for various cancers in the context of a high folate diet but not in a folate deficient diet (see text)
NAT2 N-acetyltransferase 2	Cigarette smoking	Breast cancer Bladder cancer Lung cancer Colorectal cancer	*NAT2* polymorphisms result in rapid, intermediate, and slow acetylator phenotypes. *e.g. NAT2* polymorphism encoding the slow acetylation phenotype is associated with increased bladder cancer risk amongst regular smokers
XRCC1 X-ray repair cross complementing 1	Cigarette smoking	Bladder cancer Breast cancer Lung cancer	77T>C polymorphic site in the *XRCC1* promoter region of this DNA repair gene increases the risk of NSCLC in smokers by reducing promoter activity

Diet and cancer

In 1981, two British epidemiologists, Doll and Peto, estimated that 35 per cent of cancers could be attributed to diet. The extent to which diet contributes to cancer risk varies by cancer type. The cancer most heavily implicated with diet is colorectal cancer, with as many as two-thirds of cases being attributed to an unhealthy diet, physical inactivity, and obesity. The World Cancer Research Fund/American Institute for Cancer Research (WCRF/AICR) as part of the

Continuous Update Project, continually appraise how diet, nutrition, physical activity, and weight affect cancer risk and survival using published randomized controlled trials and cohort studies. See Figure 3.4 for specific diet-related factors with convincing evidence (the strongest grade assigned) of an association with increased cancer risk.

Identifying relationships between diet and cancer from epidemiological studies is challenging, and has resulted in inconsistent findings. This is in part due to the inherent limitations of the methods used such as recall bias and self-reporting of data. Three types of epidemiological studies are typically used in diet and cancer relationship studies: case control studies, prospective cohort studies, and randomized controlled trials, and these were described in the section 'Cancer trends' at the start of this chapter.

Given that epidemiological studies can be difficult to interpret, these have evolved to include biomarker analysis as a measure of food or food group intake to validate the self-reported data. Biomarkers that can be used include carotenoids as markers for vegetable and fruit intake and urinary nitrogen excretion

Figure 3.4 Summary of strong evidence from analysis of worldwide research on diet, nutrition, and cancer.

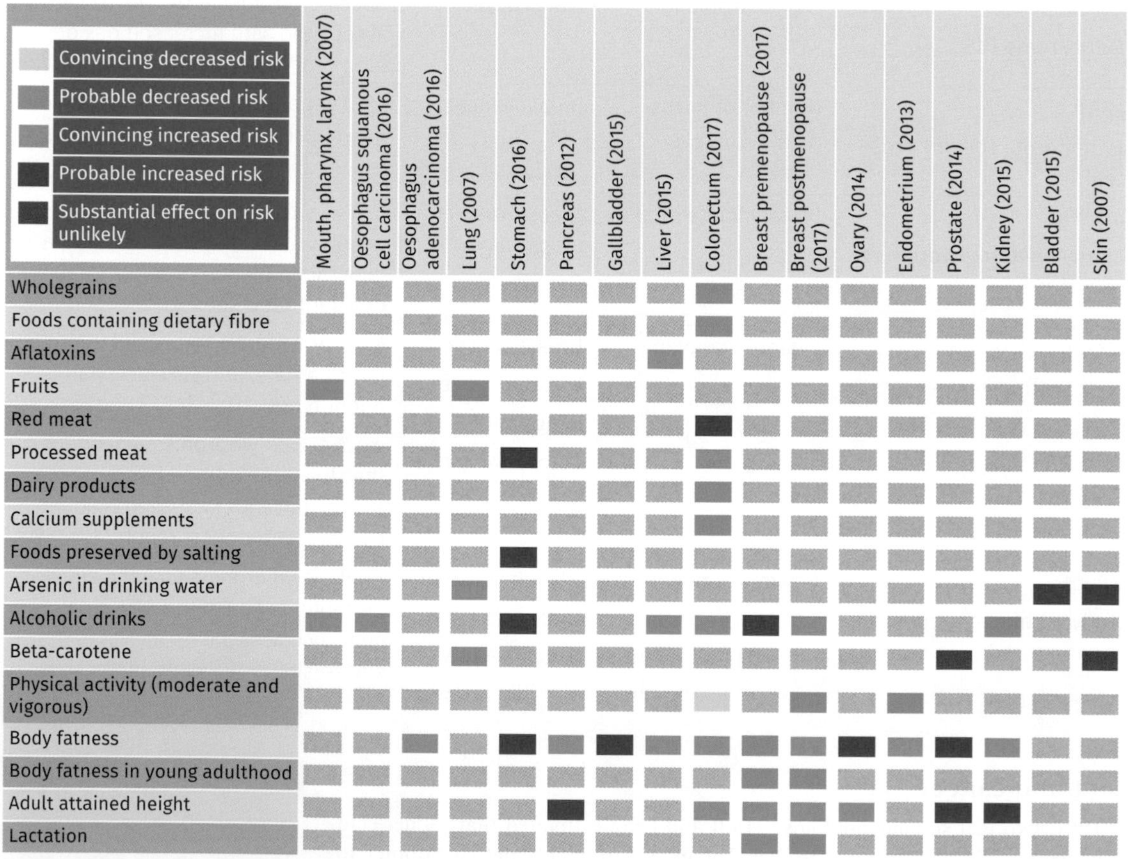

Legend:
- Convincing decreased risk
- Probable decreased risk
- Convincing increased risk
- Probable increased risk
- Substantial effect on risk unlikely

Cancer types (columns):
Mouth, pharynx, larynx (2007); Oesophagus squamous cell carcinoma (2016); Oesophagus adenocarcinoma (2016); Lung (2007); Stomach (2016); Pancreas (2012); Gallbladder (2015); Liver (2015); Colorectum (2017); Breast premenopause (2017); Breast postmenopause (2017); Ovary (2014); Endometrium (2013); Prostate (2014); Kidney (2015); Bladder (2015); Skin (2007)

Diet/nutrition factors (rows):
Wholegrains; Foods containing dietary fibre; Aflatoxins; Fruits; Red meat; Processed meat; Dairy products; Calcium supplements; Foods preserved by salting; Arsenic in drinking water; Alcoholic drinks; Beta-carotene; Physical activity (moderate and vigorous); Body fatness; Body fatness in young adulthood; Adult attained height; Lactation

Adapted from World Cancer Research Fund International/American Institute for Cancer Research. Continuous Update Project: Diet, Nutrition, Physical Activity and the Prevention of Cancer. Summary of Strong Evidence. Available at: wcrf.org/cupmatrix.

as a measure of protein intake. However, the number of biomarkers currently available for use is limited.

Cell culture and animal studies have also been used to show the effects of nutrients and other bioactive food components, and how they affect key biological processes. More recently metabolome studies have been conducted in which the full complement of metabolites within a cell have been evaluated to assess how diet and microbes in the gut (for example) may change the metabolite profile. All of these study types are included below to describe the effects of select dietary components and cancer risk, focusing on red meat consumption, alcohol intake, and obesity.

Red and processed meat consumption

Heavy intake of red meat (beef, pork, lamb, veal, mutton) and of processed meats (such as bacon, sausages, cured meats, ham, and smoked fish) is a risk factor for several cancers including colorectal and stomach cancers. The description below focuses on colorectal cancer for which the evidence for red and processed meat consumption is the strongest.

Epidemiological studies show that heavy consumption of red and processed meat convincingly increases the risk of colorectal cancer. One of these, The European Prospective Investigation into Cancer and Nutrition (EPIC) trial, published in 2005, followed 478,040 men and women from ten European countries from 1992 to1998. They found that colorectal cancer risk was positively associated with the intake of red and processed meat with a 1.35 fold risk for the high intake (>160 g/day) group compared with the lowest intake (<20 g/day).

Several mechanisms have been proposed to explain the positive association of red and processed meat consumption with colorectal cancer. Red meat contains a high concentration of haem. In the small intestine, dietary haem is degraded releasing free ferrous iron, which promotes the formation of potentially carcinogenic N-nitroso compounds (N-nitrosamines and N-nitrosamides). In addition to endogenous formation, N-nitroso compounds can also form exogenously from nitrates or nitrites added as preservatives in processed meat. N-nitroso compounds are alkylating agents and alkylation of guanine bases of DNA may lead to G to A substitution. This is a mutation common in colorectal cancers, found in codons 12 or 13 of the oncogene *K-RAS*.

Another mechanism is through the action of the potent carcinogens heterocyclic amines and PAHs. PAHs as we know are found in tobacco smoke but are also produced when meat are cooked to high temperatures, as are the heterocyclic amines. Heterocyclic amines and PAHs are not genotoxic as such, but when metabolized, their products can covalently bind to DNA mainly at guanine bases to induce DNA damage in colon epithelial cells. If not corrected, mutations can be induced which may contribute to the development of colorectal cancer.

Alcohol consumption

Heavy alcohol intake is associated with an increased risk of some cancers including mouth, larynx, pharynx, oesophageal, liver, colorectal, breast, and stomach cancers. This association is strongest with cancers of the mouth, pharynx, and oesophagus, with a relative risk in the range of 4–7 for ≥ 50 g of alcohol per day compared with none. For colorectal, liver, and breast cancer, the relative risk is approximately 1.5 for ≥ 50 g/day of alcohol consumption.

The mechanism by which alcohol causes cancer is not fully understood, and it varies by cancer type. The primary metabolite of alcohol is acetaldehyde, produced in the liver by the enzyme alcohol dehydrogenase. Acetaldehyde is carcinogenic and can bind directly to DNA forming DNA adducts. This has been demonstrated in human cells *in vitro* and in animal studies where acetaldehyde inhalation causes tumours of the respiratory tract in rats and laryngeal tumours in hamsters. Mutational signature 16 characterized by T to C substitution has been associated with alcohol consumption in oesophageal and liver cancers.

A role for acetaldehyde in contributing to carcinogenesis is also supported by findings that genetic polymorphisms in genes that code for enzymes involved in alcohol metabolism can modify alcohol-related cancer risk. The rate at which alcohol is metabolized is dependent on two classes of enzymes: alcohol dehydrogenases (ADH) which convert alcohol to acetaldehyde and aldehyde dehydrogenases (ALDH) which then convert acetaldehyde to acetate. An *ADH1B* polymorphism encoding an ADH enzyme with high activity promoting rapid conversion of alcohol to acetaldehyde appears with high frequency in Asian populations, and has been associated with an increased risk of alcohol-related oesophageal cancer. Similarly an *ALDH* gene variant resulting in the production of enzyme with reduced activity has also been associated with an increased risk of alcohol-related oral, pharyngeal, laryngeal, and oesophageal cancers in Asian populations. These *ADH* and *ALDH* variants lead to the accumulation of, and increased exposure to the carcinogenic acetaldehyde, promoting tumourigenesis.

Obesity

Obesity is associated with increased risk of several cancers including breast (in postmenopausal women), colorectal, prostate, liver, stomach, gallbladder, pancreatic, ovarian, and endometrial cancers. Obesity is defined as body mass index (BMI) of greater than 30 kg/m^2 and is a measure of body adiposity, that is, an excess accumulation of fat in adipose tissue. A large amount of epidemiological data link obesity to increased cancer risk. A systematic review and meta-analysis of seventy-six prospective cohort studies spanning North America, Europe, Australia, and the Asia Pacific involving 221 data sets and 282,137 cancer cases showed that increased BMI increases the risk for some but not all cancers. BMI-cancer associations were shown to be gender specific. For example, the risk of developing colon cancer is higher in men than in women. Although BMI-cancer associations were generally consistent across populations, some exceptions were observed. One of these was an increased risk of pre-menopausal breast cancer in Asia-Pacific populations but not in North American, Australian, and European populations.

Several biological mechanisms have been proposed to explain the link between obesity and increased cancer risk as shown in Figure 3.5. One of these is the increased production of sex hormones and applies primarily to cancers of the breast, ovary, and endometrium. For instance, in post-menopausal women, oestrogen production occurs not in the ovaries but in adipose tissue. The increased number of fat cells increases production of the enzyme aromatase responsible for converting oestrogen precursors into oestrogen. Oestrogen stimulates breast epithelial cells to proliferate and thus increases the chance of mutations occurring. Some oestrogen metabolites can also induce direct and indirect free-radical-mediated DNA damage, genetic instability, and mutations, leading to cell transformation.

A second mechanism by which obesity may cause cancer is through insulin resistance. Obesity leads to insulin resistance and correspondingly an increase

Figure 3.5 Three main biological mechanisms proposed linking obesity to tumour development. These include increased production of sex hormones, insulin resistance, and disturbed profile of adipokine production.

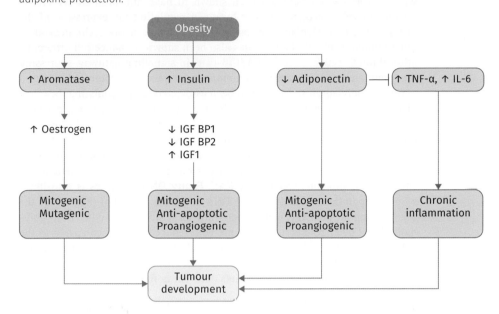

in insulin levels. Insulin promotes the production of insulin-like growth factor 1 (IGF-1) and reduces production of insulin growth factor binding proteins 1 (IGFBP1) and 2 (IGFBP2). The latter would normally bind to IGF-1 and reduce its activity. Under reduced IGFBP1/2 conditions, insulin and IGF-1 trigger signalling cascades that favour tumour development including increased cell proliferation and reduced apoptosis. This mechanism is likely in obesity-related colorectal cancer, pre- and post-menopausal breast cancer, and prostate cancer.

A third mechanism may be due to a disturbed profile of adipokine production. Adipokines are a diverse group of polypeptide hormones released from fat cells. In obese individuals, levels of the anti-inflammatory, anti-tumour adipokine, adiponectin is reduced whilst production of pro-inflammatory factors such as TNF-α and IL-6 are increased. This leads to chronic inflammation and induction of mitogenic signalling, reduced apoptosis, and increased angiogenesis, which promote tumour formation at sites such as the colon, liver, and breast.

Diet and cancer prevention

Certain food types show an inverse association, that is, a protective effect, between consumption and cancer risk. See Figure 3.4 for a summary of cancer types where an inverse association between dietary components has been established. Some of these are:

- wholegrains and foods containing dietary fibre (fruits, vegetables, and cereals) with colorectal cancer;
- non-starchy vegetables (such as carrots, broccoli, celery) and fruits with cancers of the mouth and pharynx; and
- dairy products (milk, cheese) and calcium supplements with colorectal cancer.

Nutrients within food types such as calcium in dairy products and vitamin B9 (folate) in vegetables and citrus fruits are attributed to the inverse associations with cancer risk. Calcium has been shown to have anti-proliferative effects on tumour cells, mediated in part by down-regulating the expression of the proto-oncogene c-Myc and increasing the expression of the cyclin dependent kinase inhibitor p21^{waf1}. Calcium has also been shown to induce cell differentiation through activation of the APC/β-catenin signalling pathway, supressing inappropriate β-catenin activity commonly associated with colorectal cancer.

Genetic polymorphisms have also been associated with modulating the effect of diet on cancer risk. The nutrient folate, a water-soluble vitamin of the B family is acquired through consumption of foods such as citrus fruits and dark green vegetables. Folate is required for processes such as DNA synthesis and DNA methylation. Decreased intake leads to DNA hypomethylation and inappropriate activation of genes, in particular those that carry CpG methylation sites in their promoters such as *K-RAS*. Faulty DNA synthesis is also likely with incorporation of uracil into the DNA backbone instead of thymine. In the folate metabolic pathway, methylenetetrahydrofolate reductase (*MTHFR*) polymorphism *C677T* results in an enzyme with reduced activity compared to the wild-type variant. In the context of a high folate intake, the *C677T* variant has been shown to be protective for various cancers including colorectal and breast cancer but not in individuals with diets deficient in folate.

 Key Points

- The extent to which diet and other environmental factors contributes to cancer risk is difficult to assess given the large number of variables that can confound the data. Combining data from different study-types, epidemiological and laboratory-based, can support their interpretation.
- The effect of diet and of other lifestyle factors (e.g. smoking) on cancer risk can be modified by genetic variants that an individual carries. These genetic variants function in key signalling pathways associated with the cancer phenotype.

Diet and the gut microbiome

The influence of diet on the risk of particular cancers may be due in part to changes in the microbiome. The microbiome represents the full complement of microbes in and on the human body of which the gut microbiome is the best characterized. It comprises 99 per cent of the total human microbial mass and plays an important role in a number of cellular processes in the colon, regulating epithelial cell proliferation, food metabolism, and inflammation. The number and types of microbes differ between healthy individuals and those with disease. For example, in colorectal cancer patients the numbers of *Desulfovibrio spp.*, *Enterococcus faecalis*, and *Escherichia coli* are increased, whilst the numbers of *Bifidobacterium*, *Roseburi*, and *Lactobacillus* are decreased in comparison with healthy individuals.

Indigestible food components such as complex carbohydrates (fibre), certain protein residues, and bile acids produced by the liver in response to fat ingestion are metabolized in the colon by the gut microbes. These metabolic products can be anti- or pro-carcinogenic.

Anaerobic bacteria in the gut such as *clostridales* can digest dietary fibres to produce short chain fatty acids, of which one is butyrate. Butyrate plays a beneficial role in colon health. In addition to acting as an energy source for colon cells, butyrate is anti-tumourigenic, able to suppress cell proliferation and induce apoptosis. In contrast, breakdown of proteins and fat-stimulated bile acids generate inflammatory or pro-carcinogenic metabolites. Proteins are converted to beneficial short fatty acid chains but also compounds such as ammonia, amines, and nitrates. Facultative and anaerobic bacteria are able to use the amines and nitrates to catalyse the formation of N-nitrosamines, a potent pro-carcinogen. Ammonia is usually recycled in the liver or taken up by lactobaccili and detoxified. However, at high concentrations it can induce inflammation and cell proliferation. Bile acids produced by the liver in response to fat ingestion are metabolized by sulphate-reducing bacteria and result in the production of hydrogen sulphide, a pro-inflammatory and genotoxic compound.

Studies evaluating the levels of microbial metabolites in patients with colorectal tumours compared to healthy individuals have shown that in cancer patients, metabolites such as nitrogenous compounds are increased whilst others such as butyrate are decreased. Thus a diet high in meat and fat and low in fibre, may lead to an increase in the production of pro-carcinogenic and pro-inflammatory metabolites through an altered microbial profile elevating the risk of colorectal cancer.

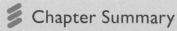

Chapter Summary

- Cancer incidence and mortality rates for different cancer types vary by geographical location, by gender, and for some cancer types, ethnicity.
- Cancer incidence rates are predicted to rise over the next fifteen years with the greatest proportional increases predicted in LMICs as they undergo economic transition and adopt the lifestyle behaviours of HICs.
- A very small proportion of cancers are due to inheritance of high penetrance genes. The vast majority of cancers are polygenic; with multiple genetic variants of moderate or low penetrance acting in combination with environmental and lifestyle risk factors to influence cancer risk.
- High throughput technologies such as next generation sequencing combined with mathematical modelling and bioinformatics have enabled distinct mutational signatures carried by cancer genomes to be identified.
- These mutational signatures provide information on the intrinsic (genetic) or extrinsic (non-genetic) factors contributing to the aetiology of the cancer.
- Diet plays a prominent role in the risk of some cancers with approximately 35 per cent of cancers linked to dietary components. Of these, consumption of red and processed meat, alcohol, and obesity feature heavily.
- The role of the microbiome in modulating cancer risk is becoming well-established for colorectal cancer.

Further Reading

Brennan, P., et al. (2011). 'Genetics of lung-cancer susceptibility'. The Lancet Oncology 12(4): 399–408.

This review provides a summary of identified genetic variants that are associated with lung cancer risk.

O'Keefe, S. J. D. (2016). 'Diet, microorganisms and their metabolites, and colon cancer'. Nature Reviews Gastroenterology & Hepatology 13: 691–706.

This article presents evidence from epidemiological, cell culture, animal, and human studies to show that a dietary imbalance has a profound effect on the gut microbiome and their ability to prevent carcinogenesis.

Petljak, M. and Alexandrov, L. B. (2016). 'Understanding mutagenesis through delineation of mutational signatures in human cancer'. Carcinogenesis 37(6): 531–40.

This paper provides an overview of the mutational signatures identified in human cancers and their future potential applications.

Sakita, J. Y., et al. (2017). 'A critical discussion on diet, genomic mutations and repair mechanisms in colon carcinogenesis'. Toxicology Letters 265: 106–16.

This paper describes the possible mechanisms by which dietary mutagens promote the formation of colorectal cancer.

Shiovitz, S. and Korde, L. A. (2015). 'Genetics of breast cancer: a topic in evolution'. Annals of Oncology 26(7): 1291–9.

This article provides an overview of high, moderate, and low penetrance genes associated with breast cancer risk.

Discussion Questions

3.1 Present a schematic showing how dietary components may contribute to colorectal cancer pathogenesis. Integrate this information with genomic mutations common in colorectal cancer and the corresponding impact on signalling pathways.

Hint: include the impact of dietary components that are tumour promoting as well as tumour protective.

3.2 Describe how mutational signatures are extracted and validated from cancer genome sequencing data. How may this information be useful in cancer prevention and treatment strategies?

Hint: remember that a cancer genome may produce a jumbled mixture of mutational patterns arising from different mutational processes and these need to be decoded.

3.3 Using a specific example of a cancer type, describe how data from epidemiological, laboratory, and human studies have informed and strengthened our understanding of cancer pathogenesis.

Hint: lung cancer could be a good example here to focus on.

4 KEY PLAYERS AND PATHWAYS IN CANCER

Learning Objectives

- define a number of interrelated signalling pathways controlled by key nodes or hubs;
- characterize the major roles of p53 in cancer;
- debate the importance of the microenvironment in tumour biology; and
- describe how cancer cells escape immune destruction.

This chapter introduces signalling pathways and their relationship to cancer. In general, tumour cells exhibit elevated cell proliferation, repressed cell death, and decreased differentiation relative to normal cells. Altered signalling in tumours was at first considered to involve linear pathways, but it is now recognized that a network view, with essential cross-talk between signalling hubs, is entailed. Thus, cancer-associated genes form key nodes or hubs in a complex of signalling pathways. As we saw in Chapter 1, activation of oncogenes such as *RAS*, *EGFR*, or *PI3K* promotes cell proliferation, whereas inactivation of tumour suppressor genes that provide cell cycle checkpoint or apoptosis control, such as *TP53*, *RB1*, *BCL2*, or *CDKN2a* (p14^{arf}, p16^{Ink4a}), prevents cell death. In this chapter, we look at the functional relationships between deregulated genes in these two groups that promote synergistic cancer signalling and tumourigenesis.

4.1 Signalling pathways

The signalling pathways that are perturbed during tumour initiation and progression underpin the hallmarks of cancer. The diversity of known changes in signalling between cancer and normal tissues is expanding and now includes changes in the tumour environment. A selection of the key pathways is described below.

Sustained proliferation

Altered expression of signal transduction pathways that stimulate cell proliferation is a hallmark of cancer. This coupled with decreased activity in tumour suppressors that normally restrict unwanted growth leads to an increase in cell mass. Existing blood vessels are inadequate for the perfusion of this new growth resulting in a hypoxic environment that favours further selective proliferation of cancer stem cells. This increase in tumour cell numbers is later supported by neo-angiogenesis and stromal cell recruitment. Thus, many tumour cell signalling pathways contribute to the promotion and regulation of sustained proliferation that eventually leads to tumour cell invasion and metastasis. Cell proliferation requires more than an increase in flux through the cell cycle. It also entails an increase in supply of energy and anabolic intermediates, in addition to decreased activity of factors that normally restrict unwanted growth. Therefore, control of proliferation is not linear but is achieved via a series of key hubs involving activation of oncogenes and inhibition of tumour suppressors.

> Angiogenesis and metastasis are described in Chapters 2 and 6, respectively.

Proliferation drivers

Hormones such as oestrogen and other growth promoting ligands, for example epidermal growth factor (EGF), stimulate proliferation by binding to their respective receptors (ER and EGFR). EGFR is a key cell surface tyrosine kinase receptor frequently mutated or overexpressed in cancer. Receptor tyrosine kinase (RTK) signalling regulates several processes involved in cell growth and survival, and RTK activation initiates signal transduction cascades resulting in DNA synthesis and cell proliferation. Figure 4.1 shows the signal transduction pathways for the proliferation driver EGF that acts as an important paradigm for the downstream effects of RTK activation.

Receptor tyrosine kinase signalling pathways are activated by ligand/growth factor binding followed by receptor dimerization and transphosphorylation. This facilitates recruitment and activation of cytoplasmic signal transducers such as Ras proteins that act as key nodes in a number of crucial pathways. Ras is activated by binding GTP catalysed by the guanine exchange factor Sos, and this in turn activates Raf, one of its main effectors. Activated Raf phosphorylates MAPKK/MEK, thus triggering the MAP kinase cascade, which finally affects the activity of transcription factors such as JUN and FOS (AP-1). Ras proteins act as key nodes for other signalling pathways including phosphatidylinositol 3-kinase (PI3K). PI3K acts through a second messenger PIP3 to activate Akt. Akt acts as a node for several functions that promote cell proliferation and survival, as shown in Figure 4.1. In normal cells, the spike of activity created by the ligand binding would then be turned off by feedback inhibition. However, these growth factor stimulated pathways are aberrantly activated in cancer and therefore targeted in tumour therapy.

> See more about targeted therapy in Chapter 5.

Figure 4.1 Receptor tyrosine kinase signalling pathways. Ligand binding causes activation of the Ras and AKT signalling hubs promoting cell proliferation and survival.

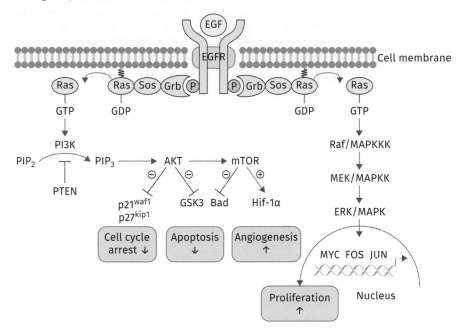

Adapted from Figure 1 in Lievre, A., Blons, H., and Laurent-Puig, P. (2010). Oncogene, 29: 3033–43.

The MAP Kinase cascade

Phosphorylation is a common mechanism for activating proteins and a series of phosphorylation steps are involved in growth signal transduction from extracellular growth ligands such as EGF to transcription factors in the nucleus. As shown in Figure 4.1, Raf is the first kinase in the cascade and this is also called MAPKKK. This phosphorylates and activates MAPKK/MEK and in turn this activates MAPK (also known as ERK). MAPK activation involves phosphorylation at both tyrosine and threonine residues. Members of the Myc family of transcription activators are also targets of MAPKs.

Loss of proliferation inhibition

Loss of tumour suppressor genes in cancers takes the brakes off proliferation without incurring concomitant cell death. The 9p21 locus codes for the cyclin dependent kinase inhibitors p15^{ink4b} (encoded by *CDKN2B*), p16^{ink4a}, and p14^{arf} (encoded by *CDKN2A*). These are vital proteins for controlling cell proliferation. Genome-wide association studies (GWAS) and gene expression analysis have demonstrated that the deletion of this locus on chromosome 9p21 is associated with an increased risk of several malignancies, e.g. melanoma and glioma. p16^{ink4a} would normally inhibit pRb, as seen in Figure 1.3. However, in the absence of p16^{ink4a}, the cell continues to cycle, unrestrained.

❯ See more about pRb and cell cycle in Chapter 1.

Induction of senescence

One of the main hallmarks of cancer is resistance or avoidance of programmed cell death, which can occur by various mechanisms. Firstly, the cell can avoid imminent cell death by inducing a state of permanent growth arrest or replicative senescence. This form of senescence, or irreversible cycle arrest, should not be confused with the temporary pause in proliferation cells undergo whilst repairing DNA. Senescence involves activation of many known tumour suppressor genes such as *RB1*, *TP53*, *CDKN2A*, and *CDKN1A*; however, it can be a double-edged sword, suppressing tumour formation in the early stages but later aiding tumour cells to avoid cell death.

Cancer cell death

A hallmark of cancer cells is their ability to evade cell death mechanisms that would lead to elimination of a tumour. The mechanisms for avoiding the various forms of cell death are described below.

Apoptosis

Apoptosis is a major pathway for regulated cell death that does not provoke an inflammatory response, as phagosomes and adjacent cells engulf cell debris. Millions of cells in our body undergo apoptosis every day and the apoptotic bodies are removed by phagocytosis. The initiation of apoptosis is controlled by the opposing actions of multiple upstream pro-apoptotic and anti-apoptotic factors.

Triggering of apoptosis can be via the intrinsic pathway involving the mitochondria and Bcl-2 family of proteins or extrinsically via activation of cell surface death receptors such as Fas by their ligands (FasL or TNF). Both pathways are shown in Figure 4.2. The intrinsic pathway is elicited by oncogene activation or other non-sustainable stresses such as DNA damage, telomere shortening, low oxygen, or growth factor availability. It is triggered when the pro-apoptotic protein Bax inserts into the outer mitochondrial membrane causing membrane permeabilization. This leads to the release of molecules such as cytochrome C and pro-caspase 9 into the cytoplasm. Cytochrome C binds to Apaf-1 in the cytoplasm to form a complex called an apoptosome and pro-caspase 9 is recruited to this complex, which leads to its activation. This is an initiator caspase, which then goes on to activate the effector caspases (caspase 3, 6, and 7) that cause cell death by cleaving vital substrates. Apoptosis is carried out by the proteolytic action of these effector caspases, and also involves DNA fragmentation by a caspase activated DNAase (CAD). Activation of CAD by destruction of its inhibitor, iCAD, results in chromatin cleavage at inter-nucleosomal linker sites resulting in the characteristic laddering effect on agarose gels.

The extrinsic pathway is triggered by binding of death initiating ligands to their cell surface receptors. Subsequently, an adaptor molecule FAS-associated death domain protein (FADD) dimerizes the death receptors, and this leads to the activation of the initiator caspase-8, thereby triggering the downstream apoptotic cascade.

Necrosis and mitotic catastrophe

An alternative method of cell death is necrosis. It is considered to be a traumatic method of cell death that releases intracellular components into the intercellular fluid where they can induce an immune reaction. Necroptosis is a programmed form of necrotic cell death that differs from apoptosis in that it involves membrane rupture and release of cytosolic contents. Like the extrinsic apoptotic pathway, necroptosis is triggered by ligand binding to death receptors such as TNFR; however, inhibition of caspase 8 activates RIPK3 and MLK1/MAP3K9 the key regulators of necroptosis. A complex called the necrosome is formed when caspase 8 is inhibited and this triggers necroptosis. Necroptosis

Figure 4.2 A schematic diagram showing intrinsic and extrinsic apoptotic pathways. The intrinsic or mitochondrial pathway is triggered by intracellular signals resulting in permeability of the mitochondrial membrane and release of cytochrome C. The extrinsic pathway is triggered by ligands binding to death receptors on the cell membrane. The two pathways converge at the terminal stages of apoptosis.

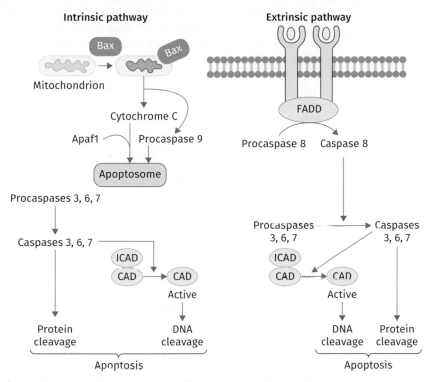

Adapted from Figure 1 in Gomez-Sintez, R., et al. (2011). Frontiers in Molecular Neuroscience, 4; article 45.

can act as a default pathway when caspase 8 activity is blocked; however, normally caspase 8 would cleave RIPK1 and apoptosis would proceed.

The term 'mitotic catastrophe' was first used by Paul Nurse to describe cells undergoing aberrant mitosis. It is associated with a defective G2 checkpoint and the formation of multinucleated giant cells. The function of the G2 checkpoint is to prevent mitosis in a cell in which the DNA is damaged or the genome incompletely replicated. ATM and ATR detect damaged DNA and then activate the checkpoint kinases, CHK2 and CHK1. These in turn inactivate cyclin dependent kinase 1 (CDK1) and cause a G2 arrest. In normal cells, this can be sustained by p53 and its target molecule p21^{waf1}. Inactivation of the G2 checkpoint in the face of sustained DNA damage results in prolonged and unwarranted CDK1 activation and cell death by mitotic catastrophe. Genotoxic signalling via CHK1/2 involves many cancer-associated genes such as *H2AX*, *53BP1*, *BRCA1*, *MDC1*, and the MRE11-RAD50-NBS1 complex. Mitotic catastrophe can also occur after damage to the mitotic spindle, for example due to the drug paclitaxel.

Autophagy

The finding that mice genetically modified to be resistant to apoptosis still underwent programmed cell death led to the discovery of an important alternative pathway called autophagy. Autophagy is a catabolic process in which damaged or non-required cellular organelles or proteins are degraded and recycled. Components marked for destruction are encapsulated into vesicles that fuse with lysosomes. Release of the enzyme lysozyme within these autophagosomes leads to the degradation of the cytoplasmic material and generation of free nucleotides, amino acids, and fatty acids, which are then reused by the cell to maintain energy production and biosynthesis. Basal levels of autophagy occur in most normal cells and are elevated under conditions of stress such as nutrient deprivation, hypoxia, or formation of reactive oxygen species (ROS), preventing damaged organelles and proteins from accumulating.

Autophagy is regulated through the activity of the protein kinase 'target of rapamycin' (mTOR). mTOR is a central regulator of cell growth, integrating signals from growth factors, nutrients, and oxygen to balance mitogenic signalling (cell survival) and cell death. Under conditions of metabolic stress, mTOR is inhibited through phosphorylation, leading to the activation of key autophagy proteins including Beclin 1 and ATGs (autophagy-related proteins). These proteins drive the formation and maturation of the autophagosome, so enhancing autophagy levels.

Impaired autophagy, like low apoptosis may contribute to carcinogenesis. Mutations in autophagy genes are found in some cancers such as glioblastoma and myeloma. Beclin1 encoded by the *BECN1* locus on chromosome 17q21 is mono-allelically deleted in several cancers such as prostate and breast cancer. Mice deficient in Beclin1 are cancer prone and Beclin1 levels are low in several cancers (breast, prostate) demonstrating a tumour suppressive role for autophagy in this case.

However, autophagy may also promote tumour progression, particularly in hypoxic conditions by furnishing malignant cells with much needed energy and recycled building blocks for growth. The tumour microenvironment is often hypoxic and therefore autophagy may act as an adaptive mechanism, allowing cells that have outgrown their vascular supply to survive by providing energy and anabolic intermediates through the lysosomal degradation of cytoplasmic constituents. Thus, the role of autophagy in cancer may be tumour suppressive or tumour promoting depending on the stage and/or type of cancer. Furthermore, there appears to be a cross-talk between the tumour suppressor p53 and autophagy. Under conditions of stress p53 is stabilized and can stimulate autophagy through mTOR inhibition. However, p53 stabilization can also trigger apoptosis, and caspases activated during this process can degrade autophagy proteins suppressing autophagy. This relationship between p53, autophagy, and apoptosis is not very well understood, but may contribute to the tumour-suppressive or tumour-permissive function of autophagy in cancer.

 Key Points

- Sustained proliferation is maintained in cancer by deregulated growth factor pathways.
- Avoidance of cell senescence or cell death in its various forms is achieved by loss of tumour suppressor function.

Genomic instability and DNA repair mechanisms

Throughout life, DNA encounters various damaging agents that cause lesions in its primary structure. As seen in Chapter 1, these lesions are detected by DNA damage pathways that sense different types of DNA damage and coordinate various cellular responses including activation of transcription, cell cycle arrest, apoptosis, and initiation of DNA repair processes. Among types of DNA

Bigger picture 4.1
The cancer cell death paradox

There is a common misconception that cancer cells are inherently resistant to apoptosis. Paradoxically, despite its renowned role in tumour suppression apoptosis frequently occurs in tumours, even at high levels where it is usually associated with poor prognosis. Indeed, a close association between high apoptotic and proliferative rates has been reported in several aggressive cancers including lymphoma, and carcinomas of lung, breast, and colorectum. Emerging evidence indicates that the effects of apoptosis can be oncogenic and even promote relapse after an initial favourable treatment response. Explanations for this lie in the all-important fact that cancer cells exist in a microenvironment of other cancer and support cells. Dying and dead cells affect this environment by producing cytokines such as VEGF and **matrix metalloproteinases** that promote neo-angiogenesis. Proliferation stimulants such as prostaglandins and other mitogenic factors released by apoptotic cells also fuel apoptosis-induced tumour cell growth. Apoptotic tumour cells stimulate tumour associated macrophage (TAM) recruitment to their environment and induce TAM proliferation. Apoptotic cells can promote tumour aggression further by biasing local immune responses towards the production of anti-inflammatory tumour promoting immune cells. Thus, high levels of apoptosis and macrophage recruitment in tumours can have an adverse effect on survival as TAMs promote angiogenesis and inhibit potential tumourcidal inflammatory effects.

❯ Antiangiogenic therapies are described in Chapter 5.

Traditionally caspases are thought of as the executioners in programmed cell death; however, recent work has revealed that they can also promote cell proliferation as part of their non-degradation functions. This dual role makes evolutionary sense, particularly when it comes to wound healing, where dying cells stimulate their own replacement. However, caspase activation in pre-malignant cells can lead to the production of mitogenic signals that stimulate excessive cell division, thus aiding malignant transformation. As a tumour grows, cells that surround it become displaced from sources of oxygen and nutrients and undergo apoptosis due to cell competition. These dying cells provide proliferative signals that further stimulate tumour cell division. It is not surprising, therefore, that high caspase expression correlates with poor prognosis for patients. This caspase-driven compensatory cell division is referred to as apoptosis-induced proliferation (AiP). Some cancer cells initiate apoptosis but fail to die and these can undergo oncogenic changes caused by the effects of caspase activated DNAase. This can promote a mutator phenotype and cancer progression. For example, glioma cells that survive the death ligand based TRAIL treatment have been shown to sustain genomic damage leading to genomic instability as a result of sub-lethal caspase activation.

Growing tumours have been likened to 'wounds that do not stop repairing', and the re-population of a tumour that occurs following therapy is an example of the normal process of tissue regeneration being hijacked in cancer. Signalling by apoptotic tumour cells may stimulate tumour regrowth for example after radiotherapy or chemotherapy. Death of the vulnerable cancer cells makes room for others that are more resistant to the treatment to proliferate and take up the vacant niche. This may select for cells resistant to apoptosis, such as mtp53 bearing cells or cancer stem cells, to fuel cancer regrowth.

Extracellular vesicles (EVs) are membrane-bound subcellular elements that contain various molecules including protein and nucleic acids from the cell from which they are derived. EVs from apoptotic tumour cells are increased in response to anti-cancer therapies and may play key roles in modifying the micro-environment in favour of tumour cell growth and survival. Apoptosis acceleration due to treatment may therefore inadvertently produce more aggressive disease. In this regard, cancer therapies designed to induce tumour cell death may sometimes be counterproductive, necessitating a re-evaluation of cancer therapy.

❓ Pause for thought

1. In what ways, could therapy-induced cell death be counter-productive?
2. Why are high levels of caspase expression associated with poor prognosis?

damage, double strand breaks (DSBs) are the most deleterious as they have a high propensity to induce genomic instability and cancer. Non-homologous end joining (NHEJ) and homologous recombination (HR) are the two major pathways responsible for the repair of DSBs. A pathway related to NHEJ is the alternative end joining (a-EJ) that employs PARP1 and unlike NHEJ, requires some degree of resection of the DNA ends. There is a further mechanism called single-strand annealing (SSA) that employs components of the other pathways. Mutations in DSB repair proteins are frequently associated with an increased risk of cancer and resistance to therapy.

Figure 4.3 shows the DNA repair mechanisms and the pathways and players involved. NHEJ, in which the ends of the DNA break can be ligated directly, often provides the first attempt at repair and is positively regulated by 53BP1. NHEJ involves PAXX-, Ku70/80-, and DNA-dependent protein kinases (DNA-PKcs), XRCC4, DNA-ligase4, and DNA polymerase. Once the protein complex is formed around the break, any damaged or mismatched nucleotides are removed, which can create mutations due to base insertions and deletions. The gap is filled in by DNA polymerase and finally ligation of the broken ends accomplished by DNA ligase IV whose activity can be stimulated by the protein XRCC4. If NHEJ is unsuccessful or if the cell is in active S phase, then HR may then be the method of choice.

DNA repair by HR is a slower but more accurate method than NHEJ. It requires a template that is provided by DNA in a homologous chromosome and involves a 'search' for a template before strand invasion and DNA synthesis in order to perform a high-fidelity repair. The key components of the HR repair

Figure 4.3 Double-strand DNA break repair pathways. The NHEJ pathway requires no homologous sequence for repair, whereas the other pathways a-EJ, SSA, and HR use some resection of the damaged DNA ends to form single-stranded overhangs and a homologous sequence as a template.

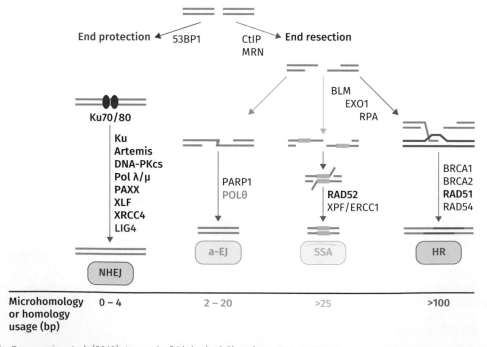

Figure 1 in Pannunzio, et al. (2018). Journal of Biological Chemistry, 293: 10512–23.

process include the proteins ATM, ATR, H2AX, RPA1, BRCA1/2, Rad51, and the MRE11-Rad50-NBS1 (MRN) complex. The procedure involves three main stages. First the DSB is recognized and processed by the MRN complex to give a 3′ single-stranded overhang. Next RAD51 binds to the single-stranded DNA, and DNA strand invasion of the homologous DNA template takes place, controlled by RAD52 and p53. The resulting heteroduplex structure is stabilized by RAD54 and p53. The role of p53 is probably to inhibit erroneous recombination thus maintaining genomic stability. Finally, the intact sister chromatid strand is utilized by DNA polymerases to re-synthesize fragments and the Holliday junctions are resolved by enzymes called resolvases.

Hypoxia signalling and cancer

Most solid tumours have regions of low oxygen (hypoxia) due to cell growth outstripping the capabilities of the local blood supply. This is aggravated by inefficient and abnormal blood vessels that are formed during tumour neo-vascularization. Hypoxic cells counteract the stress of low or variable oxygen availability by activating the transcription factor Hif-1, a master regulator of the cellular response to hypoxia. In the presence of oxygen (normoxia), Hif-1 is inactivated by the binding of the von Hippel–Lindau (VHL) protein, which leads to proteasomal degradation of Hif-1. Hypoxic conditions cause Hif-1α to heterodimerize with Hif-1β in the nucleus and in conjunction with the co-activator CBP/P300 the complex then binds to the promoters of hypoxia-inducible genes.

Hif-1 is able to activate a large array of genes involved in many of the pathways associated with the hallmarks of cancer such as proliferation, invasion, neo-vascularization, and resistance to cell death. Thus, upregulation of Hif-1, due to hypoxia or due to genetic mutations in the Hif pathway, can induce inappropriate signalling and tumourigenesis. One of the functions of Hif-1 is to switch on VEGF production leading to neovascularization of tumours facilitating their spread. More recent work has shown that Hif plays an important role in coordinating the transcription of genes involved in the control of the circadian clock. In living organisms, the circadian clock organizes the twenty-four-hour fluctuations in molecular and physiological processes that control the daily rhythm. Disruption of Hif-1 and thus disruption of circadian gene function is implicated in carcinogenesis. Pan-genomic analysis has also revealed that several cancer-risk polymorphisms are located in promoter regions of Hif target genes. These polymorphisms may therefore modulate Hif-1 signalling increasing the risk of cancer development.

Key Points

- Genomic instability and defective DNA repair provide a mutational landscape for selection of the fittest cancer phenotype.
- Hypoxia is a hallmark of the tumour microenvironment as oxygen demand outstrips supply. Upregulation of Hif stimulates neo-angiogenesis facilitating tumour growth and invasion.

4.2 p53: a master regulator in cancer

In cancer cells, key oncogenes and tumour suppressors usually act as nodal molecules that regulate signalling networks and control numerous functions that have become known as hallmarks of cancer. Some oncogene/tumour suppressor 'products' none more so than p53, satisfy the criteria of being important nodes and are critical for our understanding of carcinogenesis.

The *TP53* family

TP53, *TP63*, and *TP73* are a family of genes with overlapping as well as unique functions that play a role in cancer suppression. More is known about p53 than probably any other cancer-associated molecule and therefore a detailed description serves to highlight many aspects of cancer molecular biology. In cancer cells, p53 function can be disrupted in several ways including gene deletion or mutation. Unwarranted degradation of p53 is also a mechanism that can underpin tumour progression.

p53, the archetypal member, forms a tetrameric transcription factor that regulates many tumour suppressor hubs such as cell-cycle arrest, programmed cell death, energy metabolism, and DNA repair. In normal cells, p53 is present at very low levels unless stimulated to act as a rapid response to stress signals. A highly regulated balance between p53 production and degradation facilitates this prompt action. The low levels of p53 in normal unstressed cells are regulated by ubiquitination and proteosomal degradation involving the Mdm2-Mdm4 complex that combines with p53, marking it for destruction. The transcriptional activity of p53 is also regulated by Mdm2 and Mdm4 binding to its N-terminus. Inactivation of the p53 N-terminal transactivation domain impairs tumour suppression in mice and abolishes all p53-dependent gene expression changes, demonstrating its importance in tumour suppression.

The potent tumour suppressor effects of wtp53 are complex and affect most of the hallmarks of cancer as seen in Figure 4.4. The most significant p53 regulated pathway for tumour suppression is uncertain. For a long time, induction of apoptotic death has been regarded as the principal mechanism; however, new evidence is pointing to the possibility of metabolism. Loss of p53-driven apoptosis does not explain why p53-deficient mice are tumour-prone whilst mice lacking the downstream effectors of p53-induced apoptosis such as Bax do not develop spontaneous tumours. Interestingly, when mutations abolishing p53 regulation of metabolism-related genes are introduced into a mouse model, p53 anti-tumour functions are restrained, thereby indicating that metabolic control by p53 is crucial for tumour suppression. It should be remembered that loss of p53-induced apoptosis either alone or with loss of cell cycle arrest at G1/S does not *per se* lead to spontaneous tumour development, and the p53-directed processes and target genes critical for cancer suppression is still an open question.

p53 and the DREAM complex

p53 controls the expression of an extremely large number of genes involving both the upregulation and downregulation of mRNAs. As a transcription factor, p53 is able to bind directly to the promoter regions of its target genes to induce their expression. However, as a repressor of gene expression, p53 action is indirect, mainly attributable to effectors such as p21^{waf1} that connect p53 with the protein complex DREAM. DREAM is a transcriptional repressor comprised of members of the retinoblastoma family of tumour suppressors as well as other proteins. Its formation can occur when p21^{waf1} induced by p53, inhibits the cyclin-dependent kinase responsible for phosphorylating the retinoblastoma proteins. The DREAM complex is able to repress whole sets of genes involved in a number of cell functions including cell cycle control, DNA repair and telomere maintenance. Thus, impairment of the p53-DREAM pathway leads to deregulated cellular functions

Figure 4.4 p53 the master tumour suppressor. p53 is activated by various forms of cellular stress and then regulates most of the processes that are the hallmarks of cancer.

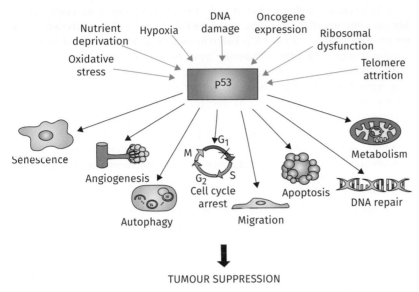

Figure 1 in Bieging, K. T. and Attardi, L. (2012). Trends in Cell Biology, 22(2): 97–106.

associated with the cancer phenotype. p53-DREAM targets include genes such as *BRCA1* and *BRCA2*, whose aberrant expression is predictive for poor clinical outcome in breast and ovarian cancers. Interestingly, the human papilloma virus (HPV) E7 protein from high-risk HPVs can abrogate DREAM regulation thus linking the p53–DREAM pathway with HPV driven tumourigenesis.

❯ See more about HPV in Chapter 1.

TP53 mutation and cancer

p53 function is drastically altered by *TP53* gene mutation, a common feature in more than 50 per cent of human cancers. *TP53* mutations are usually missense mutations and they occur mainly at approximately ten hot spots. Most missense mutations give rise to proteins that in terms of overall structure are similar to wtp53 but may be unfolded (conformational mutants) or folded but unable-to-bind-DNA proteins (contact mutants). Mutant p53 proteins can act in a dominant negative fashion. This occurs when both the wild-type and mutant alleles coexist in a cell and p53 heterotetramers with impaired transcriptional ability are formed.

The prevalence of *TP53* mutations instead of deletions suggests an oncogenic function for mtp53. Mutant p53 proteins can switch on genes that wtp53 cannot thus altering cellular functions. This gain of function (GOF) activity is achieved by mtp53 binding with a different set of transcription factors and promoters to drive transcription of a different spectrum of genes. Importantly, an intact p53 transactivation domain is required for specific mtp53 GOF actions since they are dependent on the ability of interacting proteins to partner with mtp53 and affect target gene expression.

Evidence for a gain of function effect by mutant p53 proteins has been demonstrated in murine models engineered to express different human mtp53s.

Tp53 knock-out or null mice (p53-/-), totally lacking in p53, succumb mainly to lymphomas. To demonstrate a gain of function for a mutant it must confer an enhanced oncogenic effect beyond that seen by the simple loss of p53 in null mice. In a seminal study by Olive et al. in 2004, knock-in mice with point mutations in one allele of the p53 gene at codons R172H and R270H, equivalent to the common mutations R175 and R273 seen in human cancers, showed an increased tumour burden with different tumour types compared to null mice. Moreover, p53R270H/wtp53 mice had a higher incidence of carcinomas and B-cell lymphomas compared with wtp53/- mice. In contrast, p53R172H/wtp53 mice developed aggressive osteosarcomas not seen with the R270H mutant. These mouse studies support observations in patients with Li-Fraumeni syndrome (LFS), a heritable disorder involving mutant *TP53*. LFS patients with different *TP53* missense mutations demonstrate different types and varying ages before onset of tumour formation. This clearly indicates altered oncogenic functions of mutant p53s.

TP53 and its isoforms

It has long been a mystery how p53 can carry out the innumerable functions ascribed to it, but this conundrum is now beginning to be solved. p53 function is regulated in many ways by posttranslational modifications such as ubiquitination, sumoylation, and neddylation, leading either to destruction or to phosphorylation, acetylation and methylation of p53 facilitating different cellular activities. However, another mechanism underpinning the diversity of p53 functions lies with the fact that *TP53* not only encodes a full-length protein (FLp53) but also a number of isoforms. For more than twenty-five years, FLp53 or canonical p53 (also named p53α or TAp53α) was considered to be the only isoform encoded by the human *TP53*. It is now known that at least twelve forms of p53 are produced in human cells as shown in Figure 4.5. The formation of these isoforms is achieved by transcriptional initiation from alternative promoters P1 and P2, the use of different translational initiation sites, and due to alternatively spliced pre-mRNA at intron 9 giving different carboxy-terminal domains. The intron 2 of human mRNA can also be spliced out giving an isoform with the N-terminal 40 amino acids deleted (delta40 p53/Δ40p53). Abnormal expression of the isoforms in cancer suggests they play a role, particularly in cancers that retain the wtp53. This may explain why in spite of numerous studies a clear correlation between p53 status and clinical outcomes has been difficult to establish. The p53 isoforms differentially regulate gene expression, and in normal cells this enables p53 to regulate cell fate according to the nature of a stress and cell type. In cancer cells, abnormal expression of p53 isoforms can contribute to cancer formation and progression.

Probably the best-known isoform is the amino terminally deleted form, Δ133p53. It originates from an alternative promoter site P2 within intron 4 of *TP53*, thus lacking the first 133 amino acids in humans. Δ133p53 opposes many of the functions of FLp53 and, unlike FLp53, it cannot be controlled by MDM2 mediated degradation as it lacks the relevant binding site in the N terminus. It has a significant function in stem cells where the senescence or apoptotic role of p53 would be undesirable but maintaining genome fidelity is of paramount importance. In fact, stem cell reprogramming in the absence of Δ133p53 can lead to chromosomal abnormalities.

TP63 and *TP73* isoforms and their role in cancer

The other family members, *TP63* and *TP73* can also play a role in the development and progression of cancer. Although rarely mutated, these genes have a role in cancer that is dependent on isoform expression in the context of genetic

Figure 4.5 The isoforms of p53. Full-length p53 and its twelve isoforms that are produced by alternative promoter use (P1 or P2) or truncation of the N or C-terminal regions of the protein.

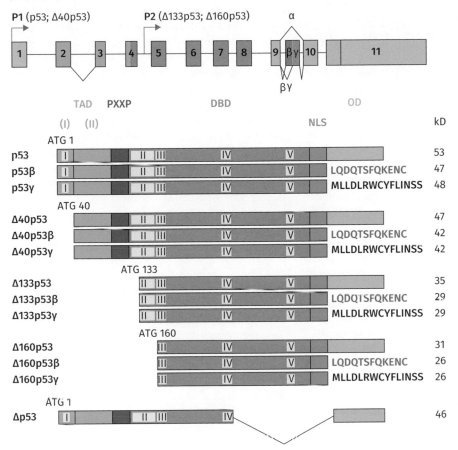

Figure 5 in Marcel, V., et al. (2010). Febs Letters, 584: 4463–68.

and cellular background. *TP63* and *TP73*, like *TP53*, also translate into several isoforms. The major isoforms are TAp63/p73 transcribed from the first promoter and Δp63/p73 transcribed from the second, internal promoter. TAp63/p73 forms are usually tumour suppressive, inhibiting tumour cell survival and metastasis. The ΔNp63/p73 isomers on the other hand have a dual role, acting sometimes as tumour suppressors but also having oncogenic properties capable of promoting certain cancers. The protein Lef1 (lymphoid enhancing binding factor 1) is a downstream player with ΔNp63 when in its oncogenic mode. These isoforms with both oncogenic and tumour-suppressor functions present a complex cross-talk with other members of the *TP53* family that is incompletely understood.

4.3 Immortality and telomere maintenance

Telomeres are found at the ends of linear chromosomes and comprise of repeats of the DNA sequence 5'TTA GGG3' with a 3'G –rich tail at the end of the repeat sequences. The complementary strand has the sequence 5'CCCTAA3' repeated. These telomeric sequences, in conjunction with associated proteins called

Case study 4.1
The p53 isoform Δ133p53β and cancer

In stem cells and pluripotent cells, p53 function has to be tempered in order for a self-renewing capacity to be maintained. The p53 isoform Δ133p53 plays a key role in maintaining this capacity and inhibiting cell death or senescence in the face of DNA damage or immune attack. Overexpression of Δ133p53 acts to inhibit the transcription of a subset of FLp53 target genes such as p21^{waf1}. Δ133p53's role in maintaining pluripotency also holds true for cancer stem cells (CSCs), the significant subpopulation of tumour cells that possess the potential for self-renewal and thus drive cancer progression.

Cancer stem cells repopulate tumours and are therefore responsible for cancer chemoresistance and metastasis formation. In breast cancer cells for example, aggressive behaviour correlates with Δ133p53β expression and an enhanced CSC population. Treatment of breast cancer cells with the cytotoxic anti-cancer drug etoposide, increases the expression of pluripotency markers such as SOX2, via Δ133p53, thus potentially increasing the risk of cancer recurrence. These findings show that Δ133p53β can support CSC potential and resistance to treatment. Thus, in this context, *TP53* can also act in an oncogenic fashion via the Δ133p53β isoform.

Δ133p53 is induced by DNA damage and antagonizes FLp53-mediated apoptosis whilst promoting angiogenesis and tumour progression. In normal cells FLp53 is selective in its triggering of DNA damage repair pathways, activating base excision repair, mismatch repair and nucleotide excision repair, and inhibiting DNA double strand break repair pathways. In contrast, the p53 isoform Δ133p53 stimulates DNA double-strand break repair protecting the cell from death or senescence in response to potentially damaging DNA double strand breaks. Δ133p53 upregulates DNA double-stranded repair by activating transcription of *RAD51*, *LIG4*, and *RAD52*. In contrast to Δ133p53, FLp53 in response to DNA damage inhibits RAD51 function. Thus, Δ133p53, an N-terminally truncated protein lacking the transactivation domain, converts p53 from a repressor to a promoter of DNA double-stand break repair thus promoting cell survival. DNA damage repair mechanisms are shown in Figure 4.3.

There is some evidence that CSCs can arise from de-differentiated and transformed somatic cells when triggered by signals from the microenvironment. Upregulation of Δ133p53 is a mechanism by which somatic cells could sustain self-renewing pluripotency whilst maintaining DNA repair capacity and resistance to senescence or apoptosis.

❭ CSCs are discussed in Chapter 1.

shelterin, cap the ends of chromosomal DNA, protecting them from erosion or end-fusions. Unprotected chromosome ends would be treated as damaged DNA, resulting in the activation of the DNA damage repair mechanism and genetic instability, senescence, or cell death. Thus, telomeres buffer the chromosome ends against loss or mutation of genomic coding sequences.

Higher organisms require larger genomes that necessitate linear rather than the restrictive circular chromosomes of prokaryotes. The downside of linear chromosomes, however, is the end replication problem. It was in 1972 that James Watson noticed that DNA polymerases could not replicate DNA all the way to the chromosome end, hence the name 'end replication problem'. Instead, the replication machinery leaves a region of the telomere uncopied and so chromosomes shorten with each round of cell division. This shortening of telomeres restricts the replicative potential of cells and is known as the 'Hayflick limit'. Telomeres shorten by at least a theoretical 20 bp each cell division, and probably as much as 100 bp due to the oxidative effects of free radicals. Telomere length declines with age from about 12 kb at birth.

Telomere shortening due the effect of ageing or chronic disease on the number of cell divisions is associated with an increased risk of cancer. In normal cells, telomere dysfunction causes activation of DNA damage responses that induce apoptosis, or cell senescence. Initially telomere shortening can be seen as tumour suppressive because it induces mechanisms to terminate further expansion of aberrantly proliferating pre-malignant cells. However, short telomeres also have the potential to promote tumourigenesis, particularly if accompanied by attenuated DNA damage responses. If telomeres become too short, they have the potential to unfold from their closed shelterin-protected structure. These uncapped telomeres can result in chromosomal fusions, leading to genomic instability and tumour cell plasticity. Thus, short telomeres, particularly in tumour initiating cells, can be oncogenic. Telomere maintenance mechanisms could then be reactivated in these transformed cells to provide immortality and ensure tumour progression.

Short telomeres in non-transformed cells of the tumour microenvironment can also affect cancer progression. Short or damaged telomeres in stromal cells can also induce senescence and the production of cytokines with protumourigenic and immunosuppressive paracrine functions. This phenotype is known as senescence associated secretory phenotype (SASP). This would allow the neighbouring tumour cell population to expand. Shortened telomeres in immune cells that impair their function might also increase cancer susceptibility.

Telomere maintenance and cancer

Cancer cells, like stem cells, must overcome the end replication problem and this is usually achieved by the action of telomerase, an enzyme capable of adding telomeric repeat sequence to the ends of chromosomes. Telomerase activity is progressively downregulated in normal somatic cells during embryogenesis but adult stem cell compartments still maintain some telomerase capability. Human tumours can originate in a telomerase positive stem/progenitor cell that already expresses active enzyme, albeit usually at a low level, or in cells that reactivate a telomere maintenance mechanism. This is usually achieved through telomerase reactivation, or a homologous recombination-based process called alternative lengthening of telomeres (ALT) in which one chromosome obtains DNA from another as shown in Figure 4.6. Activation of these telomere maintenance mechanisms is required to establish immortal growth of emerging cancer cells.

Telomerase activation

The telomere-lengthening enzyme, telomerase, was first shown in 1998 to be capable of extending cellular lifespan and soon after became recognized for its ability to immortalize human somatic cells *in vitro*. Telomerase is a reverse transcriptase comprised of a protein moiety, hTERT, and an RNA component, HTR that contains the nucleotide template for the telomeric repeats. The RNA is transcribed from the *TERC* gene. Telomerase tops up short telomeres with sufficient repeat sequences to stabilize them, allowing the overly proliferating cancer cells to become immortal. hTERT is the rate-limiting component of the enzyme, and reactivation of telomerase is a hallmark of cancer cells. The question arises: 'How does telomerase become reactivated during cellular transformation of a somatic cell?' Figure 4.6 shows some of the possible mechanisms. Re-activation of *TERT* transcription can arise as a consequence of epigenetic changes at the *TERT* promoter, gene amplifications, or rearrangements. CpG dinucleotide hypermethylation of the TERT promoter has been associated with cancer and high hTERT expression. Oncogenic mutations in the TERT promoter

Figure 4.6 Telomere maintenance mechanisms. Increased telomerase can occur by various means including (a) promoter mutation or (b) TERT gene amplification or activating rearrangement. (c) The Alternative Lengthening of Telomeres is a telomerase-independent mechanism that is favoured mainly by mesenchymal tumours.

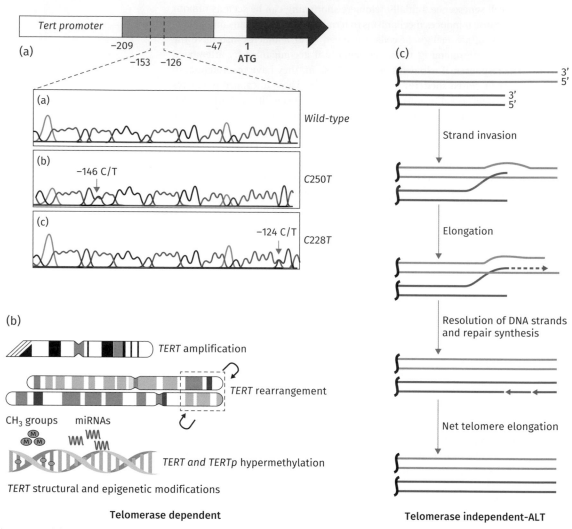

Figure 4.6(a) from Figure 5 in Giordano, M., et al. (2015). Clinical and Experimental Pathology 5: (5) DOI: 10.4172/2161-0681.1000248. Figure 4.6(b) from Figure 1 in Gaspar, B.T., et al. (2018). Genes 9: 241–99. Figure 4.6C from Figure 1 in Kass-Eisler, A. and Greider, C. W. (2000). Trends in Biochemical Sciences, 25(4): 200–4.

also facilitate gene reactivation. A causal germline mutation in the TERT promoter was initially found in a large melanoma pedigree and now *TERT* promoter mutations are possibly one of the most frequent non-coding mutations found in cancer. *TERT* promoter mutations are frequently found in tumours originating from cells with a low capacity for self-renewal whereas they are extremely rare in paediatric cancers or those from highly proliferative, stem cell rich tissues such as testicular germ cell tumours.

TERT promoter mutations often correlate with increased patient age, where the telomeres are invariably shorter. This association with telomere length could be explained by acquisition of the *TERT* promoter mutations at the point of telomere crisis during cellular transformation. h*TERT* expression is also upregulated by amplification of oncogenes such as *C-MYC* or by activation of hormone or growth factor receptors such as oestrogen receptors. Another mechanism is the loss of negative regulatory factors such as Wilms tumour suppressor1 (WT1), leading to de-repression of *TERT* transcription. Emerging evidence also indicates that the reverse transcriptase element of telomerase may have non-telomeric oncogenic functions.

Alternative lengthening of telomeres

Telomerase is active in approximately 80 per cent of tumours, whilst other cancer cells employ a recombination-based telomere maintenance mechanism called alternative lengthening of telomeres (ALT) in which one chromosome obtains DNA from another. The ALT telomere maintenance mechanism occurs in approximately 15 per cent of cancers, mainly gliomas and sarcomas. The ALT mechanism is poorly understood, but is characterized by telomeric length heterogeneity from very short to abnormally long. Another hallmark of ALT is the formation of ALT associated promyelocytic leukaemia (PML) nuclear bodies, or APBs. Replication products of this pathway such as circular C rich strands of telomeric DNA or C-circles are also present in ALT cells and along with APBs can be used to identify ALT in tumours. How ALT is activated and extends the telomere are two of the most important unresolved questions in telomere biology. Loss of the ATP dependent helicase or *ATRX* gene, usually due to truncating mutation is common, but not universal, in ALT tumours. *ATRX* knockdown in normal fibroblasts increases the proportion of cells activating ALT and accelerates the occurrence of immortalization. Interestingly, ALT is usually accompanied by mutant *TP53* in tumours. This is thought to underpin the genomic instability in ALT cells and the tolerance of defective telomeres.

Telomeres play an important role in the maintenance of genome integrity and tumour suppression and so it is not surprising that an interaction between p53 and telomere function has been described. The exact mechanism is unknown but in the presence of DNA damage, direct binding of p53 to the subtelomeric region enhances telomere stability. This may be due to the ability of p53 to recruit DNA damage repair proteins directly to subtelomeres.

 Key Points

- Cancer cell immortality requires a telomere maintenance mechanism to prevent death or senescence induction by critically short telomeres.
- Telomerase or the alternative lengthening of telomeres (ALT) mechanisms are induced in cancer cells, depending on cellular context.

4.4 Non-coding RNAs

The largest part of the genome does not code for proteins and instead gives rise to non-protein coding transcripts. Eukaryotic genomes transcribe a broad range of non-coding RNAs including ribosomal, transfer, small nuclear, and regulatory short and long non-coding RNAs.

Long non-coding RNAs

Long non-coding RNAs (lncRNAs) are RNAs over 200 nucleotides long that play a critical role in transcriptional and epigenetic gene regulation and chromatin remodelling. They play a role in lineage commitment and cellular self-renewal, and as a consequence, expression patterns have a strong tissue specificity. The human genome can code for at least 100,000 lncRNAs, some of which may even code for short proteins. They can act in cis- (e.g. lncRNA-p21) regulating expression of nearby genes or trans- altering chromatin states and gene expression in regions distant from their transcription site (e.g. HOTAIR). Although telomeres possess features of heterochromatin, they can produce a transcript TERRA (TElomere Repeat containing RNA), and p53 is required for TERRA induction in the face of DNA damage. Some lncRNAs are exported into the cytoplasm, where they can alter gene expression by affecting mRNA stability or by directly affecting protein function.

Mutation analyses of cancer genomes reveal an extensive array of functional mutations within these lncRNAs, and deregulation of lncRNA transcription also provides signals for malignant transformation. Exosome-transmitted lncRNAs released by cells function as intercellular communicators when these vesicles are taken up by recipient cells. Whilst essential for maintaining microenvironment homeostasis in normal cells, in cancers they can mediate tumour invasion, metastasis, and drug resistance. Studies have shown for example that lncRNAs exported from drug-resistant renal carcinoma cells via exosomes can transform adjacent drug-sensitive cells into a resistant phenotype.

The 3′ end sequence of *CDKN2A* encodes CDKN2BAS, an Antisense Non-coding RNA in the INK4 Locus (ANRIL). ANRIL was initially identified in familial melanoma patients and now many cancers including lung, breast, nasopharyngeal, and glioma are associated with deregulated ANRIL function. ANRIL regulates the expression of nearby genes and plays an important role in the regulation of processes such as cell proliferation, and the inflammatory response. In normal cells, E2F induces ANRIL to suppress expression of the cell cycle inhibitors of the *INK4A/B* loci in order to restart the cell cycle after transient DNA repair. In cancer cells, this mechanism is hijacked and aberrant overexpression of ANRIL prevents cell cycle halt during the DNA damage response mechanism, leading to genomic instability, and therefore, tumour progression.

Overexpression is related to cancer progression and poor prognosis since ANRIL acts by increasing cell proliferation and reprogramming energy metabolism towards increased glucose uptake and usage by aerobic glycolysis. Increased ANRIL is also reported in cancer stem cell populations. SOX2 (sex determining Y box 2) is a transcription factor essential for stem cell maintenance that binds directly to the ANRIL promoter upregulating ANRIL transcription.

MicroRNAs

miRNAs are a large class of short (nineteen to twenty-four nucleotides) non-coding RNA molecules that act as repressors of gene expression by binding to partially complementary sequences in 3′ untranslated regions (UTRs) of messenger RNA. miRNAs are involved in all the hallmarks of cancer. They can switch gene expression on or off, or fine-tune it. They are dysregulated both in tumour cells and in their surrounding stroma. Many miRNAs are tumour suppressive and target components of the cell cycle machinery such as the cyclins and cyclin dependent kinases (CDKs). A well-studied example of a tumour suppressive miRNA is miR-34a located on chromosome 1p36. miR-34a is a master regulator of tumour suppression and in normal cells it acts to promote p53 transactivation of its target genes. miR-34 functions as a significant switch in cancer development being pivotal in the decision between cell division and differentiation. Differentiation is inhibited and survival of early stage cancer stem cells promoted by low miR-34a. miR-34a is positively controlled by p53 and repressed by c-Myc. For example, inactivation of miR-34a in prostate epithelium has been shown to increase the stem cell population and lead to the development of high-grade neoplasia. In contrast, other miRNAs are considered oncogenic; the so-called oncomiRs. miR-21 was one of the first oncomiRs to be described due to its widespread overexpression in cancer and its contribution to invasion and metastases in breast cancer. Its key downstream targets are tumour suppressor genes PTEN, PDCD4, and SPRY1.

Like lncRNAs, miRNAs also play a role in communication between tumour cells and cells in the tumour microenvironment. miR-181c released from tumour cells in extracellular vesicles triggers brain metastasis by breaking down the blood brain barrier (BBB) via downregulation of 3-phosphoinositide dependent protein kinase 1 (PDPK1). Thus, secreted miRNAs represent a mechanism by which donor cells can influence physiological and pathological processes in distant recipient cells.

4.5 Metabolism

Malignant progression is accompanied by reprogramming of the cell's metabolism; this hallmark of cancer was first described by Otto Warburg. He recognized that cancer cells prefer to metabolize glucose by glycolysis, rather than through the tricarboxylic acid (TCA) cycle, even in the presence of oxygen. This may appear counter intuitive as glycolysis produces less energy per unit of glucose than complete oxidation and begs the question as to what advantage glycolysis offers over oxidative phosphorylation for cancer cells. Glucose uptake and its flux through the glycolytic pathway are increased and provide a rapid supply of energy, reducing power and biosynthetic intermediates. Glucose can be shunted into the pentose phosphate pathway to produce NADPH and precursors for the synthesis of nucleotides. Glucose also contributes to lipid biosynthesis by providing glycerol-3 phosphate and acetyl-CoA.

Tumour cells are often associated with a hypoxic environment as they outgrow their vascularity, and metastatic cells in their new environment initially lack an established vascular system and are hypoxic. Shunting large amounts of glucose through oxidative phosphorylation under these circumstances is not efficient and indeed results in excess production of radicals such as reactive oxygen species (ROS). ROS and hypoxia inducible factor (Hif-1) increase in hypoxic cells and in turn switch metabolism towards glycolysis. Thus, oxidative stress and high ROS are seen as positive accompaniments of tumour metabolism. ROS is required by cancer cells to maintain telomerase activity and in the production of invasion promoting matrix-metallo-proteinases. A feature of cancer cells is therefore adjusting metabolism to optimize intracellular ROS levels to promote growth and invasion whilst restricting apoptosis. Tumour cells also metabolize glutamine to produce the anti-oxidant glutathione as an important means of preventing detrimental ROS levels. Therefore, glucose and glutamine are important sources of energy and metabolic intermediates for tumour cell growth.

Metabolic enzymes in cancer

Although metabolic changes have been recognized as a hallmark of cancer for some time, oncogenic changes in the enzymes involved have been elucidated only relatively recently. Mutations in genes that code for enzymes of the tricarboxylic acid (TCA) cycle such as succinate dehydrogenase and fumarate hydratase are particularly important as this can lead to substantial changes in cellular metabolites. One of the most prominent examples of a mutant metabolic enzyme underpinning carcinogenesis is isocitrate dehydrogenase (IDH). The isoforms *IDH1* and IDH2 are active in the cytosol and mitochondrion, respectively. These enzymes along with NADP normally catalyse the interconversion of isocitrate and alpha ketoglutarate (αKG) by oxidative decarboxylation. *IDH1* is mutated, usually at the R132 hotspot, in cancers such as low-grade gliomas and giant cell and secondary glioblastomas. The mitochondrial IDH2 is mutated in a significant proportion of AML. Importantly *IDH* mutations are always found in the heterozygous context, i.e. mutant enzymes need to work in concert with wild-type counterparts.

The mutant forms of IDH catalyse the formation of an oncometabolite D-2-OH glutarate (abbreviated to 2HG here) by the NADPH-dependent reduction of αKG. Figure 4.7 shows the effect of mutant IDH function on several aspects of cancer related metabolism. Since 2HG is an analogue of αKG it acts as a competitive inhibitor, impeding the function of αKG dependent dioxygenase enzymes including histone demethylases and 5-methylcytosine hydroxylases such as TET that demethylate hypermethylated DNA. These are key enzymes in the epigenetic regulation of gene expression. A simultaneous increase in 2HG and decrease in αKG inhibits the chromatin-modifying enzymes that in turn inhibit expression of various genes, including those involved in differentiation. This has been shown by introducing either the oncometabolite 2HG or mutant IDH1 into normal neural stem cells and finding that the resulting epigenetic changes inhibit differentiation thus demonstrating the potential of mutant IDH to increase the chance of malignant transformation. The loss of DNA

Figure 4.7 The effect of mutant IDH on tumourigenesis. Mutant IDH1 produces the oncometabolite 2HG, a competitive analogue of αKG that inhibits αKG-dependent enzymes producing tumourigenic effects.

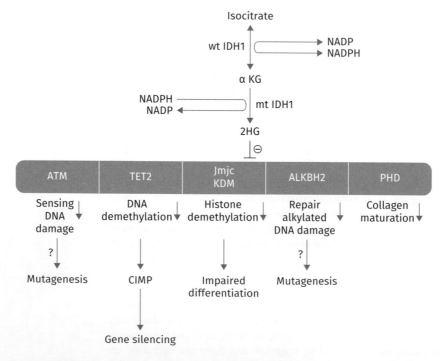

Adapted from Figure 1 in Liu, A., et al. (2016). Frontiers in Oncology, 6: article 16.

demethylation leads to a hypermethylated phenotype, mainly at CpG islands, known as a CpG island methylator phenotype or CIMP. Hypermethylation of CpG islands in promoter regions of tumour suppressor genes switches them off, promoting tumourigenesis and cancer progression.

The full effect of mtIDH and 2HG on carcinogenesis is only just being elucidated. For example, 2HG also inhibits collagen propyl 4-hydroxylase a multifunctional enzyme that can affect collagen formation. mtIDH may increase tumour mutagenesis as 2HG has been shown *in vitro* to inhibit ALKBh an enzyme that repairs DNA damage due to alkylation.

 Key Points

- Cancer cell metabolism is re-wired to support growth and cell division.
- Mutations in isocitrate dehydrogenases promotes carcinogenesis partly via the oncometabolic product 2HG.

4.6 Tumour immunity

A key hallmark of cancer is immune evasion—the ability of cancer cells to avoid destruction by the immune system. At early stages of cancer development, the immune cells are active and able to eliminate malignant cells. However, as the cancer evolves, the immune cells are 'educated' to prevent tumour destruction and so the tumour expands. This process termed immunoediting comprises three phases—elimination, equilibrium, and escape—and these are described below.

Elimination of cancer cells

Cells of the innate immune system (macrophages, dendritic cells, natural killer cells, and neutrophils) and of the adaptive immune system (T lymphocytes and B lymphocytes) play a role in eliminating cancer cells. Here we focus on the role of T lymphocytes (T cells) as a major player in this process. Naïve T cells reside in the thymus and are activated when their surface-bound T cell receptor (TCR) receives a tumour antigen from an antigen presenting cell (APC). Once activated, they migrate through the blood stream and infiltrate the tumour. Thus, T cells found at tumour sites are referred to as tumour infiltrating lymphocytes (TIL).

Tumour antigens can be of two types: tumour-associated antigens and neoantigens. Tumour-associated antigens are those that are overexpressed in tumour cells compared to normal cells. Examples include the prostate specific antigen (PSA) overexpressed in prostate cancers, HER2 in breast cancer, and mucin 1 overexpressed in several cancers including colorectal and pancreatic cancers. In contrast, neoantigens are restricted to cancer cells as they arise from somatic mutations involved in tumour formation, e.g. mutated p53 and mutated Ras.

Tumour antigens released from cancer cells are presented on the surface of APCs (dendritic cells and macrophages) through the major histocompatibility complex (MHC) cell surface receptor. Binding of MHC-antigen complexes to TCRs causes naïve T cells to differentiate into effector T cells of which one type is the helper T cell 1 (Th1) characterized by the cell surface marker CD4+. Th1 cells go on to activate another class of T cells, the cytotoxic T cells (Tc) which carry the cell surface marker CD8+. Tc cells are able to kill tumour cells via their TCR associating with MHC-tumour bound antigens presented on the surface of tumour cells. MHC molecules are of two types, class I and class II. MHC class I

Figure 4.8 Steps in the elimination of cancer cells by the immune system. Tumour antigens released by cancer cells are detected by antigen presenting cells (APC) of the innate immune system (macrophages and dendritic cells). These are captured and presented to T cells via MHC class II surface molecules in the lymph nodes. Activated T cells migrate to the tumour site, bind to tumour antigens presented via MHC class I molecules, and trigger tumour cell death. Cell death is triggered through, for example, ligands (DRL) interacting with their respective cell surface receptors (DR) located on tumour cells.

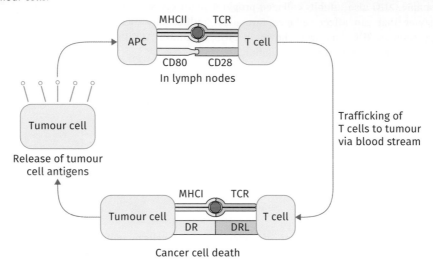

receptors are found on the surface of most cell types and present antigens to CD8$^+$ cells. MHC class II receptors are mainly restricted to the cell surface of APCs and present antigens to CD4$^+$ T cells. An overview of the steps involved in the elimination of cancer cells by immune cells is presented in Figure 4.8.

Escaping immune destruction

At the next stage of immunoediting, the equilibrium phase, immune responses against the tumour are still active but the cancer cells begin to acquire immune evasion properties. However, destruction and evasion are balanced and so there is no overall net growth of the tumour. Over time, evasion exceeds destruction and so the tumour expands entering the escape phase, the third stage of immunoediting. Cancer cells utilize a number of mechanisms to evade destruction by the immune system including:

> ### 💡 Key Points
>
> - reduction in tumour immunogenicity;
> - overexpression of immune checkpoint proteins;
> - production of immunosuppressive factors by cells of the tumour microenvironment; and
> - activation of immune cells with immunosuppressive functions such as tumour associated macrophages (TAMs) and regulatory T cells (Tregs).

Reduction in tumour immunogenicity

A key mechanism by which tumour cells avoid destruction is by downregulating the expression of MHC class I proteins, resulting in the reduction or loss of tumour antigen presentation. This reduced immunogenicity, can occur for example, through mutations in the IFN-γ signalling pathway. The cytokine IFN-γ is secreted by activated T cells and favours tumour destruction by enhancing antigen presentation on tumour cells by upregulating the expression of MHC class I receptors and by increasing recruitment of other immune cells to the tumour sites. Mutations in the IFN-γ receptor or other components of this pathway disrupt these processes and allows the tumour to escape immune destruction.

Over-expression of immune checkpoint proteins

Immune checkpoints act to regulate T cell activity, switching off T cell function when not required. This inhibition is mediated by cell surface receptors typically located on T cells binding to a corresponding ligand located on somatic cells or APCs. Two well-studied immune checkpoint proteins are CTLA-4 (cytotoxic T-lymphocyte-associated protein 4) and programmed cell death protein 1 (PD-1).

For T cells to be fully activated, two signals are needed between the T cell and the APC. The first signal is provided by the TCR binding to the MHC-antigen and the second signal by the T cell CD28 receptor binding to the costimulatory ligand CD80 or CD86 located on APCs. When T cell activity is not required, the immune checkpoint protein CTLA-4 is shuttled from intracellular vesicles within the T cell to the surface where it binds with CD80/86 preventing CD28 from binding. This interaction switches off T-cell function and downregulates the immune response. This process is shown in Figure 4.9A.

Another mechanism for switching off T cell activity when no longer required involves PD-1. PD-1 is also expressed on the surface of activated T cells and switches off T cell activity when bound to its ligand, PDL-1 or PDL-2. PDL-1 is expressed on the surface of most somatic cells when exposed to pro-inflammatory cytokines, whilst PDL-2 is located on the surface of APCs. In contrast to CTLA-4, which regulates T cell activity in the lymphoid organs at early stages of T cell maturation, PD-1 suppresses T cell function mostly in peripheral tissue and in the tumour microenvironment.

Cancer cells can exploit these inhibitory mechanisms to evade T cell-mediated destruction. For example, PDL-1 is inappropriately expressed on the surface of tumour cells in various cancer types, and on the surface of some immune cells, notably on TAMs and on Tregs.

The process by which T cell function is inhibited by PD-1/PDL-1 interaction is shown in Figure 4.9b. Briefly, activated T cells circulate throughout the body and bind to antigens expressed on the surface of cancer cells via their TCR. This binding triggers T cells to release cytokine IFN-γ, which goes onto bind to its corresponding receptor IFN-γR located on the surface of cancer cells. This interaction causes the activation of the JAK/STAT intracellular pathway leading to the expression of a number of interferon-response genes of which one is interferon regulatory factor 1 (IRF-1). Binding of IRF-1 to the promoter of PDL-1 induces its expression. PDL-1 is translocated to the surface of the tumour cell where it binds to the PD-1 receptor switching off T cell activity. Monoclonal antibodies can be directed at blocking the interaction between the immune checkpoint receptor and its cognate ligand to reactivate an immune response as a mode of therapy and is described in Chapter 5.

Figure 4.9 Inhibition of T cell activity by immune checkpoint proteins. (a) CTLA-4 inhibits T cell activation and proliferation by outcompeting CD28 for binding to costimulatory ligand CD80/86. (b) Activated T cells bind to antigens on the surface of tumour cells via their TCR. This triggers the release of IFN-γ, which through the activation of the JAK/STAT intracellular signalling cascade induces the expression of the PDL-1 ligand. PDL-1 engages with the PD-1 receptor located on the surface of T cells to inhibit its function.

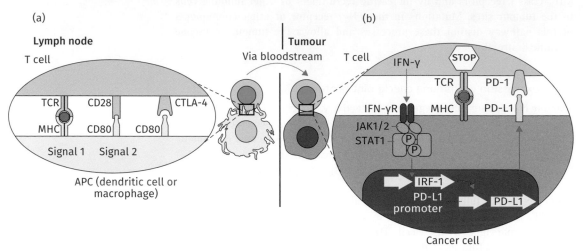

Adapted from Figures 1 and 2 in Ribas and Wolechok (2018). Science, 359: 1350–5.

Immunosuppression within the tumour microenvironment

Immune cells accumulate at the tumour site in response to pro-inflammatory molecules secreted by the tumour cells and the non-tumour cells of the TME. Pro-inflammatory molecules include cytokines such as TNF-α and IL-6, lipids such as prostaglandins and leukotrienes, and chemokines such as CCL2. These predominantly participate in the elimination of cancer cells. However, as the tumour evolves, so does the profile of inflammatory mediators moving from anti-tumorigenic to pro-tumourigenic. For example, production of VEGF, TGF-β, and IL-10 inhibits the maturation of dendritic cells decreasing their capacity to take up and present antigens to T cells. This changing profile also encourages the expansion of immune cells with immunosuppressive functions including M2 macrophages.

TAMs represent the most common immune cell type at the tumour site and are of two types, M1 and M2. M1 macrophages are able to kill cancer cells through the release of cytotoxic molecules such as TNF-α, NO, and IL-12, and are able to present tumour-associated antigens to T cells. As the tumour evolves, increased secretion of pro-tumourigenic molecules like IL-10 and TGF-β within the TME directs M1 macrophages towards the M2 phenotype. M2 macrophages support tumour progression; they are poor antigen presenters, inhibit T cell proliferation, and are able to promote angiogenesis and metastasis.

Another set of immunosuppressive cells that are activated in response to secretion of pro-tumourigenic molecules in the TME are the Tregs. These are a specialized subset of T cells characterized by the cell surface markers CD4[+], CD25[+], and FoxP3[+]. They suppress T cell activity by depleting IL-2 and preventing differentiation of T cells into effector cells. Tregs can also suppress cytotoxic T cell function via expression of immune checkpoint proteins CTLA-4 or the PD-L1 ligand.

Chapter Summary

- Mutant oncogenes form key hubs in signalling pathways that are deregulated in cancer.
- Loss of tumour suppressor genes aids avoidance of proliferation induced cell death or senescence.
- The effect of the tumour microenvironment plays an important role in carcinogenesis.
- *TP53* is the most studied gene in cancer and affects most of the hallmarks of cancer.
- Tumour cell immortality is afforded by changes in telomere maintenance involving telomerase or the alternate lengthening of telomeres (ALT) mechanisms.
- Non-coding RNAs are deregulated in cancer.
- Altered metabolism and the Warburg effect skews cellular energetics towards provision of energy and anabolic intermediates for tumour cell growth.
- At the early stages of tumour development, immune cells are able to keep the growth of the tumour in check through activation of innate and adaptive immune responses.
- However, as the tumour evolves, the TME becomes increasingly immunosuppressive and ultimately the immune responses are inhibited leading to tumour cells escaping immune destruction.

Further Reading

Allard, B., et al. (2018). 'Immuno-oncology-101: overview of major concepts and translational perspectives'. Seminars in Cancer Biology. 52: 1–11.

Summarizes the mechanisms exploited by tumour cells to escape and inhibit immune responses, and explains how these are being targeted to develop therapies.

Bourdon, J. C. (2014). 'p53 isoforms change p53 paradigm'. Molecular & Cellular Oncology 31; 1(4): e969136. doi: 10.4161/23723548.2014.969136.

A concise review on p53 isoforms and their role in modulating p53 function.

Dang, L. and Su, S. M. (2017). 'Isocitrate dehydrogenase mutation and (R)-2-hydroxyglutarate: from basic discovery to therapeutics development'. Annual Review of Biochemistry. 86: 305–31.

A detailed account of isocitrate dehydrogenases and the effects of the oncogenic mutations. The review is divided into every aspect of this enzyme's biology under a contents list so the reader can choose what is relevant for them.

Gaspar, T. B., Sá, A., Lopes, J. M., Sobrinho-Simões, M., Soares, P., and Vinagre, J. (2018). 'Telomere maintenance mechanisms in cancer'. Genes 9: 241; doi: 10.3390/genes9050241.

A comprehensive review of telomere maintenance mechanisms in cancer. This paper is extensively referenced with data collection on different cancer types and their telomere maintenance.

Koivunen, P., Sungwoo, L., Duncan, C. G., Lopez, G., Lu, G., Ramkissoon, S., et al (2012). 'Transformation by the (R)-enantiomer of 2-hydroxyglutarate linked to EGLN activation'. Nature 483: 484–8.

The definitive paper demonstrating that 2HG can stimulate Hif degradation by activating the enzyme EglN.

 ## Discussion Questions

4.1 In what ways does p53 act both in tumour suppression and as an oncogene?

Hint: think about the pathways that wild-type p53 activates to bring about tumour suppression. Think about mutant p53 and in what way it can act as an oncogene. Remember that p53 isoforms such as Δ133p53 can modulate wild-type, full-length p53 functions to subvert its suppressor actions.

4.2 One of the hallmarks of cancer is sustained proliferation. How does a tumour achieve this?

Hint: think about how growth factor pathways are deregulated in cancer and how a cancer cell avoids immediate cell death and removal. Describe ways in which proliferating cancer cells fuel a high growth rate.

4.3 How does the tumour hijack the immune system to promote its growth?

Hint: think about the mechanisms by which cancer cells evade destruction by the immune system including the generation of an immunosuppressive tumour microenvironment to gain a survival advantage.

5 CANCER TREATMENT AND CLINICAL MANAGEMENT

Learning Objectives

- evaluate various strategies for preventing life-threatening cancers;

- identify the significance of DNA damage as a means of targeting cancer cells;

- formulate an approach for treating cancer by targeting proliferation pathways;

- explain why treatments sometimes fail, including the role of cancer stem cells;

- describe how the immune system can be manipulated to treat cancer; and

- describe how clinical trials are currently undertaken.

This chapter presents an overview of cancer prognosis and current treatments such as surgery, radiotherapy, and chemotherapy. Predicting the expected outcome for patients diagnosed with cancer is a critical step in their management; however, prognostication has remained somewhat subjective, leading to suboptimal clinical outcomes. Advances in genomic and imaging technologies provide a new understanding of cancer not as a disease of an organ, but as deregulated metabolic hubs. This is paving the way to more targeted therapies. A focus on clinical trials shows how drugs are brought from the bench to the clinic.

5.1 Cancer prevention

Prevention is always better than cure and early interventions are generally more successful than later ones. This is the philosophy behind cancer screening as described in Chapter 2. Screening options are limited, however, since they require a procedure that is minimally invasive and suitable for carrying out on a large population. Tests must be highly specific, otherwise false positives can be obtained that involve the patient in invasive procedures causing anxiety and further risks. False negatives are also possible and these can lull the patient into delaying medical intervention when symptoms do appear. At present, population or mass-screening is limited to cervical, breast, and colorectal cancer.

It is estimated that at least 30 per cent of cancers are associated with known risk factors and their identification and avoidance are important for reducing cancer burden in the population. See Chapter 3 for further details of epidemiological studies that identify potential cancer risks. For example, approximately 15 per cent of all cancers are attributable to infectious agents such as *Helicobacter pylori* for gastric cancer, human papilloma virus (HPV) for cervical cancer, hepatitis B and C viruses for liver cancer, and Epstein-Barr virus for some lymphomas and nasopharyngeal cancer. Vaccines are now available for high-risk HPV variants that are being used to provide protection against sexually transmitted HPV-related cancers.

As we saw in Chapter 3, tobacco is the single greatest avoidable hazard for cancer mortality. The main tobacco carcinogens are polycyclic aromatic hydrocarbons (PAH) and nicotine-derived nitrosamines. Thus, nicotine is not only addictive but also makes a genotoxic contribution to the pathogenesis of cancer. Once activated in the cell these carcinogenic substances can form DNA adducts that cause genetic mutations or epigenetic reprogramming, leading to carcinogenic alterations such as genomic instability. Smoking for example causes higher ROS formation and consequently progressive shortening of telomeres.

Unfortunately, for many cancers such as glioma, the risk factors are unknown, hampering prevention. Identification of possible mutagens in these cancers would be highly beneficial. Since carcinogenic insults leave characteristic mutational DNA signatures, these can disclose environmental sources of mutation and thus identify hazards that may be causing cancers of unknown aetiology. Next generation sequencing is now being used to identify potential mutagenic risk factors in a variety of cancer types.

> See more about mutational signatures in Chapter 3.

Inflammatory conditions are known risk factors for several cancers, particularly colorectal cancer (CRC). Clinical studies have now provided convincing evidence that regular, low-dose aspirin use can dramatically lower the risk of CRC. Aspirin's chemopreventive properties may result, in part, from its known modulation of the inflammatory response; however, recent studies have found aspirin also alters platelet biochemistry. Platelets that normally act as first responders in wound healing also actively aid cancer growth and invasion. Aspirin may therefore be effective in cancer treatment because of its role in modulating platelet function. This is of particular interest as platelets have been implicated in aiding the metastatic process.

> See more about the metastatic process in Chapter 6.

Germline mutations in tumour suppressor genes such as *TP53*, *PTEN*, and *APC* put patients at higher risk of developing cancer. Drugs that halt or reverse the effects of tumour suppressor gene loss are being researched as possible means of cancer prevention. For example, strategies such as increasing dosage of PTEN or its activators are being explored. Research is also ongoing for drugs to block dimerization of mutant PTEN with its wild-type counterpart. Strategies for reversing loss of p53 are discussed in 5.2 below.

5.2 Cancer treatment

Surgical removal of a tumour is the mainstay of treatment for most solid cancers. Advanced minimally invasive techniques and improved anaesthesia has expanded the achievements made in this discipline. Surgery is limited by the health of the patient and the site of the tumour, for example in an eloquent part of the brain that controls critical functions. The degree of invasion of tumour cells into surrounding tissue often prevents total resection and residual disease is a major cause of recurrence. This necessitates some form of additional or adjuvant treatment, often radiotherapy either alone or with chemotherapy. Sometimes neoadjuvant treatment is given to shrink a tumour prior to surgery.

Targeting dividing cells—lethal damage to DNA or the mitotic spindle

Current standard treatments rely on targeting tumour cells by their propensity for proliferation and susceptibility to damaged DNA.

Radiotherapy

Radiotherapy is provided by high-energy X-rays that can be focused on a tumour to cause cell death or slower tumour growth. The lethal effects of radiotherapy are caused by damage to DNA in the form of base pair damage, single-strand breaks (SSBs) or double-strand breaks (DSBs). SSBs are more rapidly repaired by cells than the DSBs, which are considered to be the most lethal. One gray (Gy) of radiotherapy results in 1×10^5 ionization events per cell, producing 1,000–2,000 SSBs and 40 DSBs. A bystander effect due to irradiation of the tumour microenvironment might have an additional therapeutic benefit by removing tumour support cells.

The dose of radiation is calculated to maximize the killing effect on the tumour whilst sparing normal tissue as much as possible. This concept is called the Therapeutic Index, and reflects the comparative effects of radiation on tumour and adjacent normal tissue. Conventional radiotherapy is given in multiple smaller doses (fractionation) which also minimizes the effects on normal tissue. Early machines used magnetrons developed in the Second World War for RADAR but modern radiotherapy combines imaging with precise delivery of high-energy electron beams. Radiotherapy is given to most cancer patients as one of the most cost-effective treatments available, and contributes to a long-term remission in up to 40 per cent of patients. It is also highly effective for symptom control or palliation, especially in patients with advanced-stage cancer as it can reduce tumour burden.

Advances such as image-guided and stereotactic radiotherapy, intensity modulation, and proton beam therapy have enabled higher doses of radiotherapy to be targeted to the tumour compared to normal surrounding tissues.

External beam radiotherapy, although the most usual method of radiation delivery, is not the only one. Molecular radiotherapy introduces unsealed source radionuclides into the body, e.g. radioactive iodide to treat thyroid cancer. Radiopharmaceuticals containing a radioactive labelled drug can also be targeted to tumour cells by complexing with antibodies that bind specific cell surface proteins. Brachytherapy is a form of radiotherapy that uses a sealed radiation source placed close to the area harbouring tumour cells. Brachytherapy has been used in the treatment of cervical and skin cancer but may also be used to treat tumours in other body sites such as glioblastomas.

Oxygen increases the effectiveness of radiation by forming DNA-damaging ROS and hypoxia therefore increases radioresistance. Cells in hypoxic regions, including stem cell niches, are approximately two to threefold more radioresistant than normoxic cells. Hypoxic cell radiosensitizers such as metronidazole and hypoxic cytotoxins, such as tirapazamine, have been tried to overcome this problem. Another means of radiosensitization is to inhibit repair of the DNA damage caused by the radiation so enhancing its effectiveness. Examples of radiosensitizers used to inhibit DNA repair are inhibitors of PARP or HDAC. Cells that progress inappropriately into S phase are more sensitive to radiation, suggesting a possible role for co-administering drugs that deregulate S-phase checkpoints, e.g. fluoropyrimidines. An alternative strategy to widen the therapeutic index for radiotherapy has been to radio-protect normal tissues with approaches such as using cytokines to boost normal cell recovery, protecting with free-radical scavengers, and with stem-cell therapy.

Chemotherapy

Chemotherapy is often given in addition to the mainstays of surgery and radiotherapy. The first chemotherapeutic agents were based on mustard gas, used as a weapon in the First World War. Soldiers exposed to these agents suffered a dramatic loss of white blood cells and trials soon began to investigate the effect on leukaemia, cancer of white blood cells. Used singly, these agents gave only short-term remission but in combination with new cytotoxic drugs discovered in the 1970s and 1980s such as doxorubicin and methotrexate, transformed this lethal disease to one with long remission. Similar success was achieved for advanced testicular cancer and many childhood cancers.

Most drugs are classified according to their mechanism of action. Figure 5.1 provides a schematic of the sites of action for various types of anti-cancer drugs. For example, the nitrogen mustards (e.g. chlorambucil), nitrosoureas (e.g. BCNU), or tetrazines such as temozolomide are alkylating agents that act by covalently adding an alkyl group (R-CH2) to DNA. This inhibits enzymes involved in DNA replication and can render a cell unable to divide. Some drugs target specific parts of the cell cycle—for example, methotrexate inhibits DNA synthesis and so is most active against S-phase cells. Methotrexate is a folic acid antagonist and classified as an antimetabolite. A specific class of enzymes involved in DNA replication are the topoisomerases. They are involved in unwinding DNA and their inhibition prevents cells from entering mitosis. Examples of topoisomerase inhibitors are etoposide, irinotecan, and camptothesin. Heavy metal compounds including carboplatin and cisplatin crosslink DNA strands inhibiting replication and transcription. Cell division can also be effectively inhibited by compounds that disrupt the mitotic spindle function. Two examples of spindle poisons that have been extracted from the periwinkle are vincristine and vinblastine. Table 5.1 shows various functional categories of common chemotherapeutics.

Many forms of chemotherapy target dividing cells, the rationale being that cancer cells are more likely to be replicating than normal ones. This concept is flawed, however, resulting in serious side effects as rapidly dividing cells of the

Figure 5.1 Sites of action for anti-neoplastic drugs. Chemotherapeutic agents that target dividing cells act mainly by inhibiting DNA synthesis pathways or damaging DNA. The stages of the cell cycle where they are most effective is also shown.

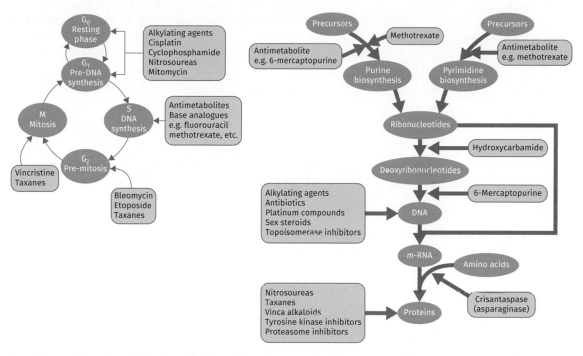

From Figures 52.1 and 52.2 in BasicMedical Key. Chemotherapy of Malignancy. https://basicmedicalkey.com/chemotherapy-of-malignancy.

gut lining, bone marrow, or hair follicles for example are lost. Cytotoxins cannot target tumour cells that are not dividing. Moreover, since cytotoxic agents target dividing cells, they have no effect on other aspects of tumour progression such as stroma modification, de-differentiation, invasion, metastasis, and development of drug resistance.

 Key Points

- Most current non-surgical interventions aim at killing dividing cells with the premise that cancer cells proliferate more than most normal counterparts.
- Radiotherapy and chemotherapy act by damaging DNA which, if unrepaired, causes cellular lethality.
- Radiotherapy and chemotherapy still provide the basis of cancer treatment.

Targeting proliferation pathways

Unlike conventional cancer treatment, which is largely cytotoxic, targeted or precision therapies act on specific molecules involved in proliferation, invasion, or tumour cell survival. As a consequence, targeted therapies are often cytostatic in that they block tumour progression rather than killing tumour cells. The key to effective precision therapy is identification of oncogene products

Table 5.1 Mode of action of chemotherapeutics

Name	Type	Mode of action	Notes
Actinomycin D	Antibiotic	Intercalates with G bases in DNA inhibiting transcription	Also a radiosensitizer as it inhibits repair of sublethal damage
Asparaginase	Enzyme	Depletes asparagine	Inhibits protein synthesis causing G1 arrest
Adriamycin	Anthracycline	Topoisomerase 2 inhibitor	Also called doxorubicin
Camptothecin	Plant alkaloid	Topoisomerase 1 inhibitor	Irenotecan is an analogue
Cisplatin	Platinum compound	Forms DNA adducts preventing cell division	Particularly effective against testicular cancer
Cyclophosphamide	Nitrogen mustard	Alkylating agent. Binds and crosslinks DNA	Suppresses the immune system. Used to treat leukaemia
Docetaxel	Taxane Anti-microtubule agent	Prevents mitotic spindle breakdown	Semi-synthetic analogue of paclitaxel
Etoposide	Plant alkaloid derivative	Topoisomerase 2 inhibitor	Semi-synthetic derivative of podophyllotoxin
Hydroxycarbamide/ hydroxyurea	Antimetabolite	Inhibits ribonucleotide reductase	Reduces production of deoxyribonucleotides and DNA
Methotrexate	Antimetabolite	Folic acid analogue. Inhibits nucleic acid synthesis	Inhibits dihydrofolate reductase (DHFR)
Tamoxifen	Non-steroidal anti-oestrogen	Steroid hormone antagonist	A prodrug activated in the liver used to treat breast cancer
Temozolomide	An imidazoltrazine, derivative of dacarbazine	Alkylating agent that binds and crosslinks DNA	Prodrug Used to treat glioma
Vincristine	Plant alkaloid from the periwinkle	Microtubule damaging agent. Prevents formation of mitotic spindle	Brand name Oncovin Used in MOPP and CHOP combination therapies

that play a pivotal role in cancer growth or survival. It is thought that some tumours rely on a single oncogene for growth and survival, in other words they are addicted to a single oncogene product. This oncogene addiction or oncogene dependency is a possible key to targeted anti-cancer therapy.

Hormone and growth factor receptors

Some tumours are stimulated to proliferate by hormones and hormone ablation is one of the oldest forms of cancer therapy. It can be achieved surgically for example by removal of ovaries to deprive breast cancer of growth promoting effects of oestrogen. Alternatively, it can be by chemical means by blocking growth factors from engaging with their receptors on tumour cells. The oestrogen blocking effects of tamoxifen for breast cancer is an example of the latter.

Many types of breast cancers require oestrogen for growth and the receptor for oestrogen (ER) is expressed in over 60 per cent of all breast tumours. Oestrogen normally binds to its receptor in the cytoplasm and the oestrogen-ER complex then translocates to the nucleus. After conformational changes, the complex binds to oestrogen response elements (ERE) upstream of oestrogen-responsive genes. These genes are mainly involved in cell proliferation and survival.

Tamoxifen blocks the action of oestrogens by binding to the ER thus inhibiting tumour growth. It is a selective oestrogen receptor modulator (SERM) and the mainstay for the treatment of hormone responsive breast cancers. Tamoxifen can also be of benefit to women at high risk of developing breast cancer.

Taken orally, tamoxifen first undergoes extensive metabolism in the liver to produce active metabolites that bind competitively to the ER, thus acting as oestrogen antagonists and inhibiting growth-promoting transcription. Progesterone receptor (PR) expression is also indirectly inhibited as it is an ER-induced gene.

Tamoxifen provides substantial improvement in the ten-year survival of women with ER-positive tumours and ER expression is currently the best predictor of treatment response. However not all patients with ER+ breast cancer respond to tamoxifen due either to initial or acquired resistance, with up to 40 per cent of patients relapsing despite anti-oestrogen treatment.

Growth factor pathways can be disabled more specifically by inhibiting ligand-receptor binding using monoclonal antibodies. Monoclonal antibodies (MoAbs) have been used for targeted therapy since the 1980s. Murine monoclonal antibodies induce an immune reaction severely limiting their repeated use, so chimeric partly humanized or fully humanized antibodies have been developed. There are several possible mechanisms of action of MoAbs including blocking signalling through receptors or inducing antibody dependent cellular cytotoxicity (ADCC). The best-known monoclonal antibody treatment is trastuzumab or Herceptin. It targets the surface antigenic sites of the growth factor receptor HER2 and is believed to block downstream signalling pathways. It can also flag tumour cells for destruction by the body's immune system, by ADCC. Almost a third of breast cancers express HER2 and these are usually more aggressive than HER2 negative. Combined Herceptin and conventional chemotherapy for patients with HER2 positive breast cancer can lead to a 50 per cent increase in survival times compared with chemotherapy alone.

Growth factor pathways are frequent targets for cancer therapy as outlined in Figure 5. 2.

Most growth factor receptors are tyrosine kinases (RTKs), enzymes that phosphorylate tyrosine residues in their target proteins. Epidermal growth factor (EGF) is a member of the growth factor superfamily that includes HER2. EGF ligand binds to the receptor EGFR thus activating its intracellular kinase domain and triggering a cascade of growth enhancing processes. Figure 5.2 shows how activation of an RTK triggers a metabolic cascade resulting in transcriptional activation and expression of proteins involved in cell proliferation and survival. The EGFR pathway is inhibited by the chimeric antibody cetuximab binding to EGFR with a similar affinity as the natural ligands, EGF and TGF-β thus blocking their effects. Small molecules targeted against the receptor tyrosine kinase (EGFR-TKIs) are also effective inhibitors of this growth factor pathway. Examples are Gefitinib (Iressa) and Erlotinib. These are used in treatment of epithelial cancers such as breast and head and neck carcinomas or

Figure 5.2 **Targeting growth factor pathways.** Drugs that act on EGFR and HER2 and their downstream signalling pathways to inhibit cell growth and proliferation. Note that HER2 has a similar downstream effector pathway to the related EGFR.

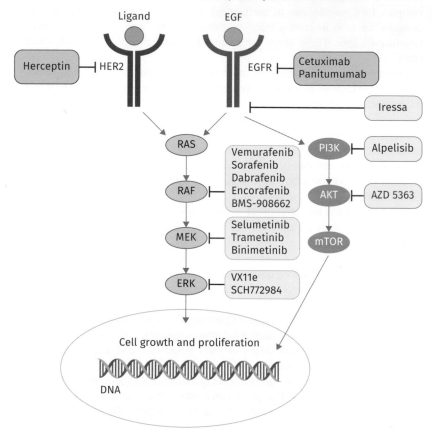

Adapted from Figure 1 in Miyamoto, Y., et al. (2017). International Journal of Molecular Sciences, 18: 752–67.

non-small cell lung cancers that overexpress EGFR or have a mutant receptor that is permanently active. TKIs are non-peptide compounds with homology to ATP that can be taken orally. They compete for the ATP binding domain in protein kinases, preventing phosphorylation and subsequent activation of downstream pathways.

Ras/GTPases

Triggering a receptor tyrosine kinase (RTK) leads to activation of the RAS proteins that form a pivotal node or hub in growth factor signalling. Briefly, upon triggering of a RTK, Ras changes its conformation as it goes from the inactive (GDP-bound) to the active (GTP-bound) form. This leads to activation of Raf serine/threonine kinases, of which there are multiple subtypes (A-, B-, and C), with different properties and functional contexts. Triggering Raf leads to activation of the downstream kinase, mitogen-activated protein kinase kinase (MEK), which in turn phosphorylates the Extracellular signal-Regulated Kinase (ERK/MAPK). After phosphorylation, ERK undergoes nuclear translocation where it activates various transcription factors including JUN and FOS, enabling expression of cell cycle proteins. The RAS pathway is among the most frequently mutated or otherwise inappropriately stimulated in human cancer and targeting drivers of this pathway with small-molecule inhibitors is a powerful treatment option for cancer patients.

❯ See more about Ras/ GTPase signalling in Chapter 4.

The first attempts to block hyperactivated RAS signalling by directly targeting mutant RAS were largely unsuccessful, possibly due to its lack of structural pockets to which small molecule inhibitors could bind to disrupt their activity. However, the B-RAF and MEK proteins, both part of the same mitogen-activated kinase (RAS-MAPK) pathway, appear to be more suitable targets. B-RAF signals to and activates the MEK kinase, leading to activation of growth signals in the cell. Thus, inhibiting downstream MEK represents another opportunity to inhibit this pathway. Targeting MEK in tumours with B-RAF mutations has been successful in melanoma patients. Mutations in B-RAF, e.g. valine to glutamic acid V600E are the most common driver mutations in melanoma and have been identified in other cancers such as colorectal, thyroid, and hairy cell leukaemia. BAY43-9006 (Sorafenib) is a small molecule inhibitor of V600E mutant B-RAF that acts by blocking the ATP binding pocket and locking the B-Raf kinase in its inactive conformation.

PI3K/AKT

There are several other effector proteins of Ras activation, including phosphatidylinositol 3-kinases (PI3K), that lead to production of phosphatidyl-inositol 3,4,5 triphosphate (PIP3). This second messenger induces a conformational change in the serine threonine kinase AKT (also known as PKB) resulting in its translocation and phosphorylation at the plasma membrane. Thus, AKT acts downstream of receptor tyrosine kinases (RTKs), such as EGFR, PDGFR (platelet derived growth factor receptor), and VEGFR (vascular endothelial growth factor receptor). Tumours harbouring deregulating mutations in RAS/MAPK and PI3K/AKT signalling require inhibition of both networks due to significant crosstalk and redundancy between the two pathways. Look again at Figure 5.2 to see where various drugs act. In fact, the PI3K/AKT pathway is another frequently disrupted signalling pathway in human cancers, where it is often associated with advanced disease and poor prognosis. AKT functions as a key node in the PI3K/AKT/mTOR pathway, influencing several cancer-related pathways as shown in Figure 4.1. Loss of the tumour suppressor gene PTEN activates the PI3K/AKT signalling pathway. Inhibitors that target AKT such as AZD5363, are now in clinical development. AZD5363 is a potent competitive kinase domain inhibitor of AKT that has been shown to inhibit tumour growth. mTOR can also be targeted by a category of drugs called rapalogs, which have shown some success in clinical trials.

Synthetic lethality

Synthetic lethality is one of the most important new concepts for drug discovery. A synthetic lethal interaction occurs between two genes when the alteration of either gene alone allows cell viability but when altered simultaneously results in cell death. Cancer therapies can exploit synthetic lethality where a drug that inhibits the protein product of a gene is effective in combination with mutation of a second gene. Loss of tumour suppressor gene function is common in cancer, but it can be tolerated if an alternative pathway is exploited. Synthetic lethality occurs when a gene from this second pathway is also deleted or its product inhibited. This should reduce toxicity since only cells with the mutated gene, cancer cells, should succumb to the drug treatment. For example, the cell adhesion gene for E-cadherin (*CDH1*) is frequently mutated in cancer and this loss of E-cadherin imparts a dependence on the receptor tyrosine kinase, ROS1. This dependence of mutant *CDH1* on ROS1 can be exploited by drugs such as foretinib and crizotinib that inhibit the latter. ROS1 inhibition is therefore a synthetic lethal with E-cadherin deficiency. Another example is synthetic lethality between mutations in *BRCA1* and PARP1 inhibitors in breast cancer. PARP1 functions by modifying nuclear proteins involved in DNA repair. It is important for cell recovery after DNA damage, particularly when BRCA function fails. PARP1 inhibitors are therefore particularly effective in tumours that have lost BRCA function due to their ability to induce synthetic lethality.

Targeting oncoproteins

The key to successful cancer therapeutics is to target selectively the genetic lesion that initiates and maintains the proliferative potential and survival of a cancer cell. Some cancers are sensitive to inhibition of a single oncogene, a concept referred to as 'oncogene addiction'. The Bcr-abl fusion oncoprotein produced by the Philadelphia chromosome (Ph) is both necessary and sufficient to drive chronic myeloid leukaemia (CML) and is an example of 'oncogene addiction'. CML is a haematological stem cell malignancy with a reciprocal chromosome translocation (t[9:22] [q34:q11]) called the Philadelphia chromosome (Ph). The breakpoint cluster region (Bcr) on chromosome 22 and the Abelson kinase on chromosome 9 give rise to the oncogenic fusion protein Bcr-abl. Bcr-abl is a constitutively active tyrosine kinase that affects several downstream pathways such as MAPK, PI3K, STAT5, and Myc. This results in increased cell proliferation and defective apoptosis leading to an expansion of malignant pluripotent cells in the bone marrow. This is shown diagrammatically in Figure 5.3. CML progresses through three distinct clinical stages: chronic phase, accelerated phase, and blastic stage. This last stage is the most aggressive and characterized by a high number of blast cells in the blood and bone marrow. Imatinib is a clinically approved tyrosine kinase-inhibitor that specifically blocks the binding of ATP to Bcr-abl, thus inhibiting its kinase activity. Imatinib is the highly specific, first-line drug of choice for the treatment of patients with chronic phase CML, inducing long lasting remissions with few side effects. Other tyrosine kinase inhibitors developed for CML include Dasatinib and Nilotinib. Unfortunately, some CML patients develop resistance to Imatinib due to mechanisms such as Bcr-abl amplification or mutations. Mutations can cause direct steric hindrance for binding of the drug to the active site of the enzyme, and are thought to be a main reason for treatment failure. Bcr-abl can also promote genetic instability by down regulating DNA repair enzymes leading to an increased mutation rate. A mutation at threonine 315 (T315I) renders CML resistant to Imatinib and is the most common mutant in Imatinib resistant CML. This resistance has been specifically overcome by the development of the new generation inhibitor Ponatanib.

TP53

As seen in Chapter 4, *TP53* is the most frequently disabled gene in human cancer and therefore restoring wild-type p53 function or inhibiting mutant p53 would appear to be an obvious potential targeted treatment. In addition to abrogating wild-type p53 functions, mutant p53 can also act through its gain of function (GOF) properties.

Figure 5.3 The Bcr-abl signalling network. A diagram of the chimeric Bcr-abl oncoprotein that has constitutive kinase activity. (a) Bcr-abl is retained in the cytoplasm where it activates cell proliferation and survival pathways. (b) The main site of action of the Bcr-abl inhibitors are shown.

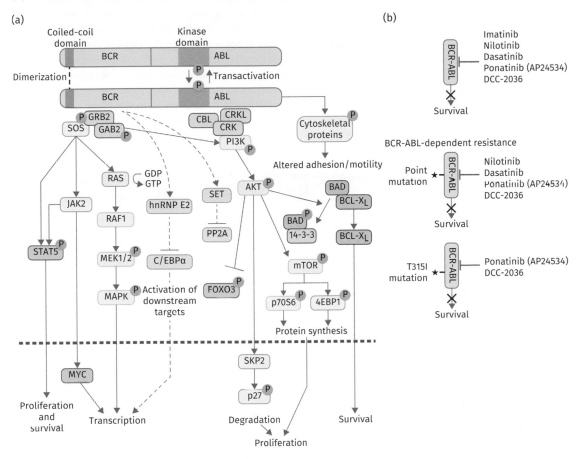

Adapted from Figure 1 in O'Hare et al (2010). Clinical Cancer Research, 17(2): 212–21.

Most *TP53* mutations are missense, occurring mainly in a few hotspots, particularly at arginines R175, R248, R273, and R213. Many mutations lead to conformational changes in p53 and small molecules have been developed that can re-establish the wild type conformation. One example is the PRIMA family of small molecules that have been shown to restore mtp53 proteins to a wild-type conformation *in vitro*, re-establishing some wtp53 transcriptional activity such as enhanced expression of Puma, Noxa, and Bax in p53 mutant cells.

One gain of function property of mtp53 lies in its ability to bind to p63 and p73 and inhibit their activity. The binding of mtp53 to p73 can be inhibited by the compound RETRA, leading to restoration of p73-dependent transcription and cell death. High levels of mtp53 depend on the chaperone HSP90 for transactivation of target genes and inhibition of HSP90 either alone or in combination with histone deacetylase (HDAC) inhibition has been shown to have anti-tumour activity in mice. However, due to the complexity of this signalling hub, most anti-mtp53 compounds have been found to show off-target activities and have failed to make it to the clinic. Development of drugs to target the actions of mtp53 has been hindered by the lack of understanding of the molecular mechanisms underpinning reactivation of wild-type p53 function in mtp53 cells.

Wild-type p53 cells repair damaged DNA at G1 but mutant p53 cells in the absence of a functional p53 dependent G1 arrest, rely more on a G2 checkpoint for repair. Therefore, targeting the G2 checkpoint could inhibit DNA repair in mtp53 cells causing apoptosis or mitotic catastrophe. Inhibitors of the G2 checkpoint can sensitize mtp53 tumour cells to DNA-damaging agents. This is demonstrated in mtp53 colorectal carcinoma cells, in which the preclinical WEE1 kinase inhibitor, PD0166285, potentiated radiation-induced killing. This caused loss of G2/M arrest thus forcing premature mitotic entry, mitotic catastrophe, and cell death. The G2/M regulator WEE1 is also highly expressed in several other cancer types such as non-small cell lung cancer and melanoma making it a possible target for disarming mtp53 cancer.

Key Points

- New targeted therapies have been developed that disrupt growth factor signalling pathways.
- Designer drugs that specifically inhibit oncoproteins such as the Bcr-abl kinase provide remissions with very few side effects.

Targeting neo-angiogenesis

Vessel recruitment by tumour cells is a hallmark of cancer that could be an important target for therapy. However, as shown in Table 5.2 the mechanisms for vessel recruitment in tumours is varied, complicating the development of robust antiangiogenic therapy. Solid tumours are characterized by high levels of proangiogenic factors such as VEGF, FGF, and PDGF that activate signalling pathways stimulating microvascular proliferation. Antiangiogenic strategies have focused on targeting these pathways. Thalidomide is a weak inhibitor of VEGF- and FGF-mediated angiogenesis, and is a first-line treatment in multiple myeloma. This is a cancer of plasma cells for which bone marrow vascularity is a marker of poor prognosis. The success of thalidomide in treating multiple myeloma may also depend on non-anti-angiogenic actions of this drug such as its immunomodulatory, anti-proliferative, and pro-apoptotic properties.

Anti-angiogenic strategies have focused more recently on the use of antibodies that bind VEGF such as bevacizumab (Avastin), a humanized monoclonal antibody. Ramucirumab is a fully humanized monoclonal that binds VEGFR2. Drawbacks to the use of bevacizumab include its immunogenicity and potential to induce autoimmune disease after prolonged treatment. It would be desirable if small molecule tyrosine kinase inhibitors that block receptor activation

Table 5.2 Mechanisms of vessel recruitment

1) Neo-angiogenesis, a process whereby new vessels sprout from existing ones.
2) Angiogenesis from bifurcation of existing vessels.
3) Vascular co-option, tumour cells migrate and hijack existing vessels.
4) Vascular mimicry, a process whereby new blood vessels are formed without endothelial cells but are lined by tumour cells instead.
5) Cancer stem cells differentiate into endothelial cells.

could be used to avoid these long-term limitations of antibodies. Early generation small VEGFR inhibitors such as sunitinib lacked specificity and inhibited many kinases with similar potency. It has, however, been shown to have efficacy in treating renal cell carcinoma (RCC). Fruquintinib, a new generation small molecule inhibitor, has high and selective potency for VEGFR kinases and inhibits tumour growth through its antiangiogenic properties. Ziv-aflibercept acts as a decoy receptor that binds VEGF ligands decreasing their effect on native receptors. In addition to VEGF function, non-VEGF mechanisms of blood vessel formation have also been investigated as antiangiogenic targets including Delta-like 4/Notch and TGF-β type 1 receptors. A schematic of the anti-angiogenic drugs targeting these pathways is shown in Figure 5.4.

Antiangiogenic therapies tend to be cytostatic rather than cytotoxic and sustained dosage is therefore important. Tumours acquire a blood supply through several mechanisms and drug resistance may result as alternative routes to angiogenesis are acquired. On the downside, suppressing neo-angiogenesis may select for aggressive tumour cells that adapt to a hypoxic environment. For example, in GBM antiangiogenic therapy induces transformation from a proneural to a more invasive mesenchymal phenotype.

Figure 5.4 Angiogenesis inhibitors as anti-cancer drugs. Inhibition of tumour angiogenesis by blocking the proangiogenic signaling pathways that would lead to proliferation of tumour vasculature.

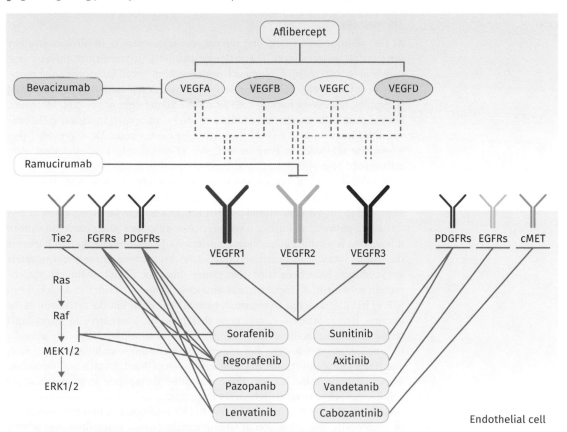

Figure 1 in Comunanza V and Bussolino F (2017). Frontiers in Cell & Developmental Biology, 5: 101.

If a cancer is driven by loss of a tumour suppressor gene causing abnormally increased expression of proteins, the therapeutic approach is to target its downstream pathway. An example is the von Hippel-Lindau tumour suppressor gene and renal cancer. Clear cell renal cell cancer (CCRCC), is associated with functional loss of the von Hippel-Lindau tumour suppressor gene leading to elevated levels of Hif-1α and consequent overexpression of VEGF, increased tumour angiogenesis, and hypervascularity. This maybe an example of another form of 'oncogene' addiction as RCC is driven by VEGF overexpression.

Treatment with anti-angiogenic agents also changes gene expression in stromal cells, such as cancer associated fibroblasts (CAFs), as they adapt to hypoxic stress. For example, an isoform of VEGF, VEGF-α is upregulated in CAFs under anti-angiogenic drug treatment in proportion to the severity of resultant hypoxia. Angiopoietin-2, which is expressed in endothelial cells is similarly upregulated and can act as a Tie2 receptor agonist promoting sprouting angiogenesis and therefore limiting the effect of VEGF inhibition. Thus, stromal cells can secrete hypoxia-stimulated signalling molecules that contribute to tumour angiogenesis, and hence to anti-angiogenic drug resistance.

Targeting telomeres and their maintenance

One of the hallmarks of cancer is a capacity for unlimited replication; as such, it is a potential target for cancer therapy. Two mechanisms, telomerase and ALT, are known to maintain telomere length and thus support immortalization of tumour cells.

❯ See more about telomeres and their maintenance in Chapter 4.

Telomerase

At first sight, it seems targeting the enzyme telomerase is an obvious strategy for treating most cancers. The rationale behind anti-telomerase therapy is it may lead to a decrease of telomere length resulting in cell senescence and apoptosis of tumour cells. Telomerase activity provides replicative immortality and is elevated relative to normal cells in almost 90 per cent of cancers. Moreover, normal stem cells usually have longer telomeres compared to cancer cells, providing at least some degree of specificity for cancer cells. Thus, provided that telomerase inhibition is time limited, side effects due to loss of normal stem cells should be minimal.

In normal cells, a certain threshold of telomere attrition and associated reduction in shelterin activates the damage-repair system as it recognizes the unprotected DNA double strand as DNA breaks and activates the p16^{ink4a} or p53 signalling pathway to initiate a senescence or apoptosis programme. In cancer, this system is disabled often through deletion of chromosome 9p that harbours the *INK4A* locus or by mutation of *TP53*. Several strategies for direct telomerase inhibition have been tried, that target either the hTERT or the ribonucleoprotein subunit hTR. These include antisense oligonucleotides to knock-down hTR or hTERT transcript levels, small molecules that inhibit the active site of the enzyme, immunotherapies, and gene therapies. An alternative to direct inhibition of the enzymatic activity of telomerase is to inhibit its ability actually to function at the telomere. For example, G-quadruplex stabilizers, tankyrase, and HSP90 inhibitors target the telomere structure itself, frustrating telomerase assembly there. Table 5.3 lists these and other therapeutic strategies that are being approached to target telomeres in cancer.

Unfortunately, it requires only a few hTERT molecules to prevent senescence in cancer cells, and inhibition of telomere maintenance is required over a large

Table 5.3 Therapeutic approaches to telomerase inhibition

1. **Oligonucleotide inhibitors**

    a) RNAi based TERT knockdown has been used experimentally.

    b) Oligonucleotide drugs use antisense technology to inhibit hTR.

    c) GRN163L (Imetelstat) is a modified oligonucleotide template antagonist that binds with high affinity to hTR and blocks interactions between telomerase and telomeric DNA.

2. **Small-molecule telomerase inhibitors**

    a) Azidothymidine (AZT), which is a thymidine nucleoside analogue, can be effective in targeting the active site of TERT. It acts as a reverse transcription inhibitor.

    b) BIBR1532 a small non-nucleosidic synthetic compound that inhibits the processivity of telomerase *in vitro*.

    c) A series of imidazol-4-one derivatives are reported to inhibit telomerase.

3. **Immunotherapeutic approaches**

    Immunotherapeutic approaches directed against TERT epitopes involving peptides derived from TERT—for example, the GV1001 telomerase vaccine.

4. **Telomerase-directed gene therapy**

    Telomerase-directed suicide gene therapy using a prodrug that is activated in telomerase positive cells.

5. **G-quadruplex stabilizers**

    G-quadruplex structures form in the G-rich 3′ strand of telomeric DNA.

    Telomestatin (SOT-095) is a compound that interacts with and stabilizes G-rich configurations. This blocks interactions between telomerase and telomeric DNA, preventing telomerase from functioning there.

number of population doublings before it induces tumour suppressive senescence. This necessary long-term hTERT inhibition may be deleterious in highly proliferative telomerase dependent cells as telomeres become damaged or dysfunctional long before cell death is induced. In the meantime, before cell death or senescence is induced very short and damaged telomeres could promote tumour growth due to chromosomal instability. This is exemplified in dyskeratosis congenita, a rare disorder characterized by short telomeres in which critically short telomeres unable to bind the protective shelterin complex result in genomic instability. Shorter telomeres are also associated with factors of poor prognosis such as higher interleukin-6 (IL-6) and C-reactive protein levels. These limitations may explain the moderate success rates in many clinical studies. Moreover, hTERT inhibition may also interfere with its telomere-independent physiological functions such as DNA repair or ribosome biogenesis.

Although direct inhibitors of telomerase have not been overwhelmingly successful in the clinic, many clinically approved drugs have been identified that appear to show inhibition of telomerase activity as an additional or off-target effect. These include molecules which downregulate hTERT gene transcription such as tyrosine kinase inhibitors, ubiquitin/proteasome inhibitors, and cytotoxic drugs such as temozolomide (TMZ). TMZ inhibits hTERT transcription via the Sp1 transcription factor binding site. Other drugs act via hTR downregulation such as DNA topoisomerase1 inhibitors and cisplatin. Interestingly, the circadian rhythm hormone melatonin downregulates both hTERT and hTR at a transcriptional level. Several pitfalls in telomere targeting in cancer are listed in Table 5.4.

Table 5.4 Pitfalls for telomerase inhibition

1. Since the anti-proliferative effects of telomerase inhibitors are only induced in cells with short telomeres, there is a requirement for several tumour cell divisions before the drug can be effective.
2. Telomerase inhibition may elicit negative effects in highly proliferative normal cells, such as stem cells, causing side effects.
3. Telomerase inhibition may cause senescence and senescent cells may go into crisis, producing pro-tumourigenic paracrine effects.
4. Induction of telomere dysfunction and attrition induces chromosomal instability that may increase the number of activated oncogenes and loss of tumour suppressor genes which promote malignant transformation and drug resistance.
5. hTERT inhibitors may exhibit 'off target' effects that promote the selection of tumour-promoting properties including chromosomal instability.
6. Telomerase inhibition may select for tumour cells that can use the ALT mechanisms of telomere maintenance.
7. Tumours that have the ALT mechanism would be unresponsive at best.

Alternative Lengthening of Telomeres

A significant proportion of tumours naturally maintain their telomeres through the alternative lengthening of telomeres (ALT) mechanism as described in Chapter 4. Sarcomas have a high frequency of ALT where it is a prognostic factor for poor outcome in patients. It is also possible that the use of telomerase inhibitors in tumours heterogeneous for a telomere maintenance mechanism will select for clones that use the ALT mechanism and cause the failure of these drugs as therapy. ALT mechanisms are still poorly understood and measurement of ALT is difficult, hampering the development of a necessary readout for measuring drug efficacy. Moreover, an ALT-like mechanism has been found in some normal cells. However, long and heterogeneous telomere lengths, a characteristic of tumour ALT, is not seen in normal cells so this maybe a valuable tumour target. Trabectedin is an alkaloid derived from a sea squirt that was originally found to have therapeutic benefit in sarcomas. Trabectedin interacts with the minor groove of the DNA helix where it alkylates guanines, triggering interference with transcription factors, DNA binding proteins, and repair pathways, resulting in cell cycle arrest and apoptosis. ALT positive cells have been shown *in vitro* to be more sensitive than ALT negative ones to trabectedin.

Drugs that target telomere replication in general have been shown to be anti-proliferative *in vitro*. Long ALT telomeres are rich in G-quadruplexes and ligands that bind to these G-rich repeats could be suitable drugs to target ALT tumours. Drugs that specifically target telomeric G-quadruplex such as telomestatin have been shown *in vitro* to inhibit the telomeric homologous recombinational basis of ALT with no or low toxicity in normal cells. The consequences were shortened telomeres and ultimate apoptosis or senescence. ALT cancers should be sensitive to inhibitors of the ALT homologous recombination process such as the cisplatin derivative, Tetra-Pt(bpy). Unlike cisplatin, Tetra-Pt(bpy) does not form covalent DNA adducts and has a limited effect on the proliferation of normal cells. It should therefore have fewer side effects than cisplatin. However, Tetra-Pt(bpy) might have a slower action than cisplatin, as there is a lag time between drug administration and cell death being triggered by extremely short telomeres.

Key Points

- Telomerase inhibitors are now in clinical trials that target telomere maintenance and tumour immortality.
- The ALT pathway at present is poorly understood with few means of measuring outcomes, which hampers drug design and clinical trials.

Targeting developmental pathways and cancer stem cells

There is overwhelming evidence in support of the hypothesis that malignant tumours are initiated and maintained by cancer stem cells (CSCs). Some CSCs can arise by de-differentiation of a more differentiated cancer cell that then acquires self-renewal properties and clonal repopulation potential. CSCs have a high propensity for self-renewal but a slow growth rate, and therefore are resistant to therapies that target dividing cells. CSCs also escape from immune surveillance. This and their survival in the face of DNA damage results in renewal of cancer tissue leading to treatment failure and tumour recurrence. Targeting CSCs is a priority, but a major hurdle that needs to be overcome is to find drugs that specifically target malignant stem cells, keeping normal stem cells intact. In addition, treatment with CSC-specific drugs alone would not suffice since it would leave the bulk of tumour cells intact and give no relief from symptoms associated with tumour burden.

Cancer stem cells are being targeted by monoclonal antibodies to cell surface markers, for example anti-CD44 for treating acute myeloid leukaemia and anti-CD133 for liver cancer. Alternatively, these surface markers can be exploited as a means of specific drug delivery to the cancer stem cells. For example, nano-particles loaded with paclitaxel for the treatment of CD44 expressing breast cancer. Differentiation therapy is another strategy to target CSCs as in the treatment of acute promyelocytic leukaemia (APL). All-trans-retinoic acid and arsenic trioxide cause degradation of the oncogenic retinoic acid receptor alpha (RARα) that is the underlying oncogenic deficit in these cells, leading to leukaemic cell terminal differentiation and cell death. A remission in the majority of APL patients is achieved by a combination of cytotoxic and differentiation therapy.

As tumorigenesis like embryogenesis depends on coordinated mechanisms of proliferation, differentiation and migration both have several functions in common. Signalling pathways involving developmental genes such as Notch and Hedgehog (Hh) are the hallmarks of CSCs. These developmental pathways are also implicated in regulation of EMT, increased invasion, and metastatic potential.

Targeting Hh or Notch and eradication of CSCs is therefore of prime importance for tackling cancer. The ligand for the hedgehog pathway, Hh, binds the receptor patched (Ptch) and activates the G protein smoothened (SMO) as shown in Figure 5.5. This in turn stimulates the functional transcriptional activators of the Hh pathway, the glioma-associated oncogene homologues (Glis). Hh/Gli play a central role in tumour cell survival from DNA damage induced by chemotherapy or radiotherapy. There is new evidence that Hh signalling can co-operate with other relevant cancer associated pathways such as EGFR, RAS, TGF-β, Notch, and PDGFR in a context-dependent manner. Stimulation of the EGFR/RAS/MEK pathway can activate Gli transcription in different cancer cell lines from brain, prostate, pancreas, and breast cancers. Targeting Gli with inhibitors

❯ See more about cancer stem cells in Chapter 1.

❯ See more about EMT and metastasis in Chapter 6.

Figure 5.5 Hedgehog signalling and its inhibition in cancer. Inactive Hh is shown on the left and activated signalling on the right with sites of action of inhibitors. Shh = Sonic hedgehog.

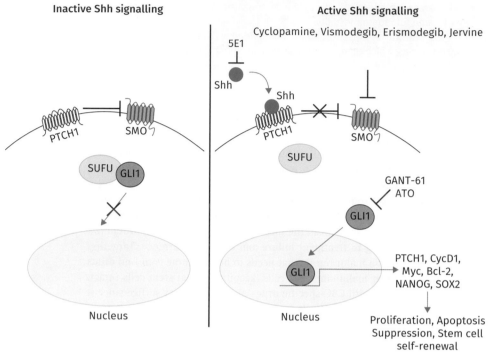

Adapted from Figure 1 in Rimkus, T. K., et al. (2016). *Cancers*, 8: 22–45.

such as GANT61 and arsenic trioxide (ATO) is an attractive proposition since Gli can be activated by both Hh ligand dependent and independent pathways.

Inactivating mutations of the tumour suppressor PTCH are the primary cause of the abnormal activation of the Hedgehog (Hh) pathway and upregulation of SMO in basal cell carcinoma (BCC) of the skin. BCC is the most commonly diagnosed cancer and inhibitors of the Hh pathway have been developed for its treatment. SMO inhibitors such as cyclopamine or vismodegib applied topically usually cause tumour regression; however, some tumours do develop resistance. Activating mutations in SMO that maintain Hh signalling in the presence of SMO inhibitors have been identified in many resistant BCCs. Developmental gene signalling such as Hh enhances DNA damage repair and upregulation of Hh pathway function can affect the response of gliomas to the DNA damaging drug, TMZ. The Gli inhibitor, GANT61, has been shown to enhance sensitivity to TMZ in glioma cells. Inhibition of Hh pathway may therefore offer a possible target to increase sensitivity to DNA damaging drugs. Look at Figure 5.5 to see how the Hh/Gli pathway is being targeted.

Targeting the epigenome

Disruption of normal epigenetic regulation is a hallmark of cancer and may be amongst the earliest changes in tumourigenesis. Genes affecting epigenetic regulation are often deregulated in cancer, emphasizing a role for therapeutic targeting of epigenetics. Moreover, unlike genetic mutations, epigenetic marks are reversible, increasing the chance of ameliorating the disease phenotype.

❯ See an introduction to epigenetics in Chapter 1.

The fundamental epigenetic mechanisms operating in cancer cells include DNA methylation, post-translational modifications of histone proteins, and mutations in chromatin remodelling proteins. The addition of a methyl group to the cytosine residues is catalysed by DNA methyltransferases (DNMTs). The epigenetic drugs azacytidine and decitabine inhibit DNMTs, inducing DNA hypomethylation and reactivation of silenced genes leading to cell differentiation and tumour suppression. Other nucleoside analogues that similarly target DNMTs are now in pre-clinical or clinical trials. DNA methylation is a recognized mechanism associated with chemotherapeutic resistance in cancer cells, and targeting the epigenome with DNA methylation inhibitors may restore expression of silenced genes and re-sensitize resistant tumours to chemotherapeutics.

Histone modification by acetylation changes chromatin structure and thus plays a key role in the epigenetic regulation of gene expression. Acetylation of histones and other proteins is maintained by a balance in the activity of the enzymes histone acetyltransferases (HATs) that add acetyl groups and histone deacetylates (HDACs) that remove them. HDACs also regulate the acetylation status of a variety of other non-histone substrates, including key tumour suppressor proteins and oncogenes. Deregulation of HDAC genes is linked to tumour development through modification of gene transcription and also of non-histone HDAC substrates.

High expression of several HDACs is associated with cancers such as prostate, gastric, lung, oesophageal, colorectal, and breast. Inhibition of HDACs is therefore one of the most promising approaches to targeting the epigenome in cancer. Valproic acid (VPA) has been prescribed for many years as an antiepileptic drug, however, it was later recognized to have cancer inhibitory activities through its action as a histone deacetylase inhibitor. Two other HDAC inhibitors (HDACis) that have been clinically approved are SAHA (vorinostat) for the treatment of advanced cutaneous T cell lymphoma and FK228 (romidepsin) for the treatment of peripheral T cell lymphoma. HDAC inhibitors have shown limited effect as single agents, but can be used in combination with kinase inhibitors, autophagy inhibitors, or ionizing radiation. For example, VPA increases the efficacy of radiotherapy and temozolomide in glioblastoma patients. VPA decreases some DNA repair processes that would otherwise protect cells against DNA-damaging agents. Thus, HDAC inhibitors may be useful as chemosensitizers to increase the efficacy of chemotherapeutic compounds. HDAC inhibitors have also been shown to reduce angiogenesis and modulate some immune responses.

HDACs are necessary for the maintenance of genes that are essential for survival and growth of cancer cells, which seem to have a vulnerability where epigenetics is involved. For example, HDACis suppress the DNA repair proteins, MRE11 and RAD50, but only in the cancer cells, leading to the relative specificity of HDAC inhibition. However, HDACis are not without effect on some normal cells and the antiproliferative activity of HDACis maybe dosage and context dependent. For example, the HDACi VPA inhibits invasion in bladder cancer but not in prostate cancer cells.

HDACis—a cautionary tale

Some HDACis can act as inducers of cancer stem cell properties and activate EMT, promoting invasion and metastasis. Thus, although VPA can induce the differentiation of some cancer cells *in vitro* and suppress tumour growth

in vivo, other studies have suggested that VPA can also amplify and maintain CSCs in certain systems. For example, HDACis may act as a potent regulator of CD133, a transmembrane protein that is highly expressed in pluripotent
and CSCs. CD133+ve cancer stem cells show a higher resistance to anti-cancer
treatments than CD133–ve cells. VPA has also caused enhanced EMT in colorectal cancer cells and SAHA, a broad spectrum HDACi with epigenetic activity
promotes EMT in triple negative breast cancer cells via HDAC8/FOXA1 signalling. Further research is obviously needed to evaluate the possibility of EMT or
CSC induction by epigenetic therapy prior to the clinical use of HDAC inhibitors
and demethylating agents.

 Key Points

- Cancer stem/progenitor cells are targets for new drug development as they
  are highly resistant to conventional therapies causing relapse and tumour
  progression.
- Cancer is a disease of epigenetics as well as genetics and targeting the epigenome is an attractive proposition for new drug development.

5.3 Immunotherapy

Immunotherapy aims to treat cancer by generating or enhancing an immune
response against the tumour. It was named by the journal *Science* as the
breakthrough of the year in 2013 and, more recently, the 2018 Nobel Prize in
Physiology or Medicine was awarded to the two scientists for their efforts in
developing this mode of therapy. Immunotherapy differs from other methods of
cancer treatment in that it does not target the tumour cell directly but instead
targets the immune system. Principally three strategies are utilized: immune
checkpoint blockade, adoptive T cell transfer, and cancer vaccines.

Blocking immune checkpoints to enhance T cell response

T cell activity is regulated by receptor-ligand pairs called immune checkpoint
proteins. One of these pairs is the CTLA-4 receptor, which binds to the ligand
CD86 or CD80 on APC cells, thus switching off T cell activity. Similarly, the PD-1
receptor is also inhibitory down-regulating T cell activity once bound to its cognate ligand PDL-1 expressed on the surface of most somatic cells and PDL-2 present on APCs. These mechanisms ensure that T cells are active only when required.
Cancer cells exploit these checkpoints to evade immune destruction. The pro-inflammatory environment of the tumour microenvironment can lead to inappropriate expression of checkpoint proteins: CTLA-4 and PDL-1 are expressed on the
surface of the immunosuppressive Tregs and PDL-1 is expressed on the surface of
tumour cells. CTLA-4 competes for binding with the co-stimulatory ligand CD86
or CD80 on APCs whilst PDL-1 binds to the PD-1 receptor on T cells. Both interactions suppress T cell function, thus preventing elimination of tumour cells.

❯ See an introduction
to tumour immunity in
Chapter 4.

Monoclonal antibodies can be directed to these checkpoint proteins to prevent receptor-ligand interactions and thus restore T cell activity as shown in
Figure 5.6. Ipilimumab blocks the CTLA-4 receptor from binding to its corresponding ligands, and was the first such therapy to be approved in 2011 for the
treatment of metastatic melanoma. Subsequently monoclonal antibodies pembrolizumab and nivolumab, directed against the PD-1 receptor, were approved

Figure 5.6 Immune checkpoint blockade. (a) CTLA-4 binding to co-stimulatory ligand CD80 can be blocked by anti-CTLA-4 monoclonal antibodies. (b) Interaction of the PD-1 receptor with ligand PDL-1 can be blocked by directing monoclonal antibodies to the receptor (anti-PD-1 antibodies) or to the ligand (anti-PDL-1 antibodies).

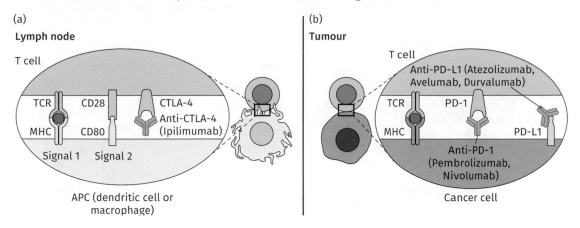

Adapted from Figures 1 and 2 in Ribas and Wolechok (2018). Science, 359: 1350–5.

in 2015, again for metastatic melanoma. These have been followed by approval of antibodies asterolizumab and durvalumab targeting the PDL-1 ligand in 2016 and 2017, respectively, for a small subset of metastatic cancers.

Although monoclonal antibodies directed against checkpoint proteins are better tolerated and have fewer side effects compared to chemotherapy, they are still associated with substantial inflammatory effects. These immune related adverse events (irAE) mainly involve the gut, skin, liver, lungs, and the endocrine glands but can potentially affect any tissue. In addition, treatment benefits are observed only in 20–30 per cent of patients in most tumour types. Targeting multiple checkpoints using a combination of checkpoint inhibitors show greater effectiveness than targeting a single checkpoint but the side effects are correspondingly increased compared to monotherapies.

Other checkpoint regulators that restrict the ability of T cells to kill tumours effectively have also been discovered. These include TIM-3, LAG-3, VISTA, and TIGIT. Antibodies against these checkpoint proteins are under development, with many showing effectiveness in pre-clinical testing.

Adoptive T cell transfer with a specific focus on CAR T cells

In adoptive T cell therapy, T lymphocytes are harvested from a patient, manipulated *in vitro*, and then infused back into the patient. These manipulations may include expansion of tumour-reactive T cells or more commonly genetic modification of the T cells to carry a synthetic TCR (sTCR) or a chimeric antigen receptor (CAR). The aim of these manipulations is to generate T cells that are able to induce a strong immune response against the tumour cells when returned back into the patient.

CARs are transmembrane proteins comprised of an extracellular antigen recognition domain and an intracellular T cell signalling domain, separated by a transmembrane region as shown in Figure 5.7. The antigen recognition domain is a single chain variable fragment (scFv) derived from an antibody targeting a specific tumour antigen. In the first CARs produced, the T cell signalling domain

Figure 5.7 Chimeric antigen receptor (CAR) designs. (a) First generation CAR comprised of a single chain variable region (ScFv) of an antibody (antigen binding site), a transmembrane domain, and a T cell activation domain (CD3ζ). (b) Second generation CAR includes a costimulatory domain (CD28). (c) Third generation CARs include two co-stimulatory domains derived from different co-stimulatory proteins. (d) Fourth generation CARs enhanced to induce anti-tumour effect include TRUCKS or armoured CARS (see text).

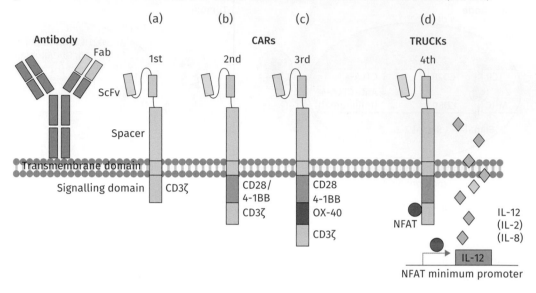

Adapted from Figure 1 in Kalaitsidou, M., Kueberuwa, G., Schütt, A., and Gilham, D.E. (2015). CAR T-cell therapy: Toxicity and the relevance of preclinical models. Immunotherapy, 7(5): 487–97.

comprised of the CD3ζ portion of the TCR. Since then, improved second and third generation CARs have been developed that include additional stimulatory domains derived commonly from co-stimulatory molecules CD28 or 4-1BB, or both, to enhance T cell activation.

CAR T cells are assembled in the laboratory by transferring the CAR construct encoded as a single gene made up of component domains into the genome of T cells using a replication-incompetent γ-retrovirus or lentivirus. Once infused back into the patient, CAR T cells expressing the chimeric receptor bind to tumour antigens on the surface of cancer cells without the need for presentation via the MHC proteins. This therapy has been used most successfully against haematological cancers. In August 2015, the FDA approved the first CAR T cell therapy for use with patients aged up to twenty-five years with refractory and/or relapsed B cell acute lymphoblastic leukaemia (B-ALL). Tisagenlecleucel, or Kymiriah™ as it is commonly known, is directed against the antigen CD19, which is expressed on B cell cancers and on differentiated B cells but not on haematopoeitic stem cells or any other essential cell types, thus limiting autoimmune toxicity. However, CAR T cell therapy is associated with cytokine release syndrome (CRS), which is characterized by a massive release of a variety of cytokines by the activated T cells and other immune cells such as the macrophages. CRS can lead to fever, hypotension, and neurotoxicity in the patient.

Using CAR T cell therapy in solid tumours is more challenging than in haematological malignancies since the immunosuppressive tumour microenvironment can dampen T cell function. To address this challenge, fourth-generation CAR T cells are being designed. One strategy is to include an inducible cytokine gene cassette to drive the expression of pro-inflammatory cytokines such as IL-12. These TRUCKS (T cells directed for universal cytokine killing), secrete IL-12 upon CAR engagement with the antigen, recruiting endogenous immune cells to the tumour site and enhancing their anti-tumour effect. A second strategy is to generate armoured CARS so that T cells no longer express immune check-point proteins such as CTLA-4 or PD-1.

Another challenge to using CARs in solid tumours is selecting an appropriate antigen to target. Most targets selected to date are tumour-associated antigens that are expressed at high levels in tumour cells but also in healthy cells although at lower levels. Examples of selected targets include prostate-specific membrane antigen (PSMA) for prostate cancers, HER2 for HER2+ tumours, and mesothelin for pancreatic and ovarian cancers. Given the non-specificity of the selected target antigens, the potential for irAEs is also high. A major thrust of research is now to identify and target neoantigens thus personalizing CAR T cell therapy.

> The effect of the microenvironment on immune cell function is discussed in Chapter 4.

Cancer vaccines

Cancer vaccines can be prophylactic or therapeutic. The aim of prophylactic vaccines is to prevent cancer formation and these vaccines are therefore effective in patients who have not yet developed cancer. In contrast, the aim of therapeutic vaccines is to treat established disease. Both vaccine types require identification of an appropriate antigen against which the body can mount an immune response. This has been possible for cancers that are caused by viruses and two such vaccines have been approved for use, Gardasil and Cervarix, directed against the human papilloma virus.

Gardasil is a quadrivalent vaccine that confers protection against the four HPV types 6, 11, 16, and 18, whilst Cervarix is a bivalent vaccine providing protection against HPV types 16 and 18. Seventy per cent of cases of cervical cancer worldwide are associated with HPV-16 and -18, and 90 per cent of cases of external genital warts in men and women are estimated to be caused by HPV-6 and -11. Hence Gardasil can confer protection against cervical cancer and genital warts. Both vaccines are prepared from the L1 structural proteins of the HPV types against which it provides protection and have been shown to prevent the development of precancerous lesions (CIN) caused by HPV-16 and -18 with near 100 per cent efficacy when administered in previously uninfected individuals.

The vast majority of cancers are non-infectious and therefore appropriate antigens cannot be easily identified. In the case of therapeutic vaccines, as the cancer is already established, it is possible to identify tumour-associated antigens or neo-antigens against which an immune response can be invoked. Therapeutic vaccines are at early stages of development.

> See more on therapeutic vaccines in Chapter 6.

5.4 Clinical trials

New candidate drugs require rigorous testing for safety and efficacy and must be an improvement on standard care and practice if they are to be approved for use in the clinic. Phase 1 clinical trials are undertaken to establish safety and side effects of a drug around its presumed optimum dose. Phase 2 trials aim to confirm the efficacy of a drug against its target cancer. Drugs that still look promising then go on to stage 3 trials, in which they are compared blind with standard current treatments.

The pre-clinical phase of drug development identifies suitable compounds based on molecular targets and testing efficacy *in vitro* on cancer cell lines or *in vivo* in animal models. Further *in vitro* toxicity testing is carried out to determine if drug levels that could be tolerated by patients would provide a dose sufficient to kill cancer cells. Thus, escalating doses are given to animal models to see if appropriate levels are safely achievable. Having established a candidate drug that satisfies pre-clinical trials, phase 1 trials of potential cancer drugs are carried out in small numbers of patients who have exhausted standard options. Ever-increasing doses are given to establish side effects of the drug until the maximum tolerated dose has been established. The disadvantage of these trials is that they are carried out on patients who are desperate for a drug with therapeutic benefit and who by the very nature of phase 1 trials stand a low chance of receiving an effective dose. Moreover, these patients have usually exhausted standard therapies and often have developed drug-resistant tumours. In phase 2 trials, on the other hand, the risk of under- or over-dosing is reduced and fitter patients are selected on the basis of likely benefit compared to the risk. Disease response is measured based on reduced tumour load, which can be complete or partial, or, failing that, whether stable disease can be achieved. Drugs that show promise in stage 2 trials go on stage 3, where they are compared in random trials with current standard regimes. Since differences in survival between trial arms is often small, large numbers of patients need to be recruited with medium to long-term monitoring, making these expensive trials. It is now increasingly recognized that patient selection for stage 3 trials should be based more on molecular defects than on organ location.

A major reason for radiotherapy and conventional chemotherapy remaining the mainstays of cancer treatment is due not only to the high cost of bringing new drugs into the clinic but also a failure to improve patient outcomes significantly in clinical trials.

5.5 Therapeutic problems

Single agent therapies can have excellent initial efficacy but rarely give sustained responses as resistance develops due to rewiring of signal transduction pathways and clonal evolution. Therefore, combination therapy is usually more effective, and investigating combination therapy for combating cancer resistance is currently of great interest in the clinical setting. The combined effect of two drugs can be synergistic, additive, or antagonistic. This depends on whether the combination gives greater, equal, or less than the sum of the individual effects of the drugs. Thus, antagonism does not mean that one drug totally cancels out the efficacy of the other but that the additional benefit from the second drug is smaller than in the additive case. Synergism is intuitively attractive, as it may allow for lower individual drug dosage; however, it may not be the best choice since therapeutic success ultimately depends on delayed development of resistance. Cancers are heterogeneous and drug resistance is a process of mutation and evolutionary selection of resistant subclones. If cells arise that are resistant to one drug, they will escape both the effect of that drug, and any synergistic enhancement it conferred to the second drug. Thus, unless

Case study 5.1
Glioma subtypes, prognosis, and treatment

High-grade glioma or glioblastoma multiforme (GBM) is the most common primary tumour of the brain. The current therapeutic regime for newly diagnosed GBM is surgical resection followed by radiotherapy and adjuvant therapy with the alkylating drug temozolomide (TMZ). Despite this aggressive therapy survival remains on average twelve to fifteen months, with very few longer-term survivors. Evidence based therapy for glioma is urgently needed.

Recent advances made in molecular profiling techniques have affected the classification of glioma subtypes. In 2016, the World Health Organization introduced IDH mutational status into the **CNS** tumour classification. Wild-type IDH1 gliomas are usually more aggressive and occur in older patients than their mutant IDH1 counterparts. Wild-type IDH1 gliomas produce large amounts of lactate via glycolysis (Warburg effect) that causes a lowering of intercellular pH and death of neurons and other brain cells. This augments infiltration of the tumour cells into brain parenchyma and thus tumour expansion. Lactate also triggers an immune response, recruiting macrophages to the tumour site. These tumour-associated macrophages (TAMs) produce cytokines and growth factors that drive tumourigenesis. In addition, the low pH environment has a detrimental effect on normal lymphocytes that perform immune surveillance.

❯ See an introduction to TAMs in Chapter 4.

IDH mutant gliomas occur most commonly in patients between the ages of twenty and forty. They are more likely to originate in the frontal lobes and they have lower levels of necrosis compared to the wild-type IDH gliomas. Their reduced invasive potential compared to wild-type IDH tumours usually makes a more complete surgical resection possible. IDH mutations drive cell transformation primarily due to the epigenetic modifications caused by 2HG that blocks cell differentiation and facilitates increased cell proliferation. Beside these epigenetic modifications, 2HG has also been shown to regulate the activity of several enzymes as shown in Figure CS5.1. For example, D-2HG produced by mutant IDH1 has been observed to activate the Hif propyl hydroxylase, EGLN, leading to Hif-1α degradation. Recent data have also shown that 2HG decreases mTOR signalling which in turn would reduce Hif-1α and VEGF levels as shown in Figure 4.1. These low levels of mTOR and Hif-1α in mutant IDH gliomas suppress some of the cancer hallmarks associated with aggressive behaviour such as angiogenesis and fits with their comparative more indolent nature.

However, although intrinsically less aggressive than their wild-type counterparts, mutant IDH1 gliomas are nevertheless usually fatal. Mutation of IDH is an early oncogenic event in these gliomas and their overall metabolism reflects an adaptation to this change. For example, the reduced availability of NADH for countering oxidative damage means that alternative mechanisms for reducing the effects of ROS have been employed. The metabolic adaptation to a mutant IDH environment also includes frequent selection of a mutant p53 and loss of the chromatin remodelling protein, ATRX. This may increase tolerance of genomic instability and also hamper therapeutic response to DNA-damaging therapies. Thus, it has been shown that radiation treatment results in a more rapid repair of DNA breaks in mutant IDH than wild-type IDH tumour cells, which could signify a reduced therapeutic effect of DNA damage. Pathways involved in DNA repair such as homologous recombination and base excision repair are enriched in mutant IDH cells. On the other hand, IDH mutant cells are highly dependent on Bcl-xL and Bcl-2 for survival and Bcl-2 inhibitors such as ABT-199 may have potential for targeting these tumours.

IDH mutation is an attractive therapeutic target since it is an early change, homogenously expressed and persistent in tumour cells. Efforts to translate this genetic lesion into effective treatment have been hampered by the difficulty of measuring **objective drug responses** in clinical trials. Repeated tumour sampling by biopsy is not an option and a surrogate marker of response is needed. Moreover, since the blood brain barrier is less compromised in mtIDH gliomas sampling body fluids is particularly unreliable for these tumours. An alternative possibility is provided by the oncometabolite 2HG produced by mtIDH. 2HG is both an attractive biomarker of mtIDH activity in a tumour and also a potential target in clinical trials. Imaging of 2HG has been shown to be a surrogate for measuring mtIDH tumour burden in patients and therefore may be a valuable 'read-out' for efficacy in clinical trials. However, although targeting mtIDH may reduce 2HG levels this may not equate to tumour cell death and imaging 2HG may sometimes give a pseudo-response. This is a situation analogous to the effects of antiangiogenic therapy in glioma patients when tumours become almost invisible on MRI due to the reduced vascularity resulting in a reduction in the contrast enhancement and not a clinical response of the tumour.

Figure CS 5.1 Targeting mutant IDH1 and IDH2 in cancer. Secondary and lower grade gliomas usually harbour mtIDH1 and occasionally mtIDH2. A schematic of the changes in metabolism caused by these neomorphic enzymes and the means of targeting them is shown.

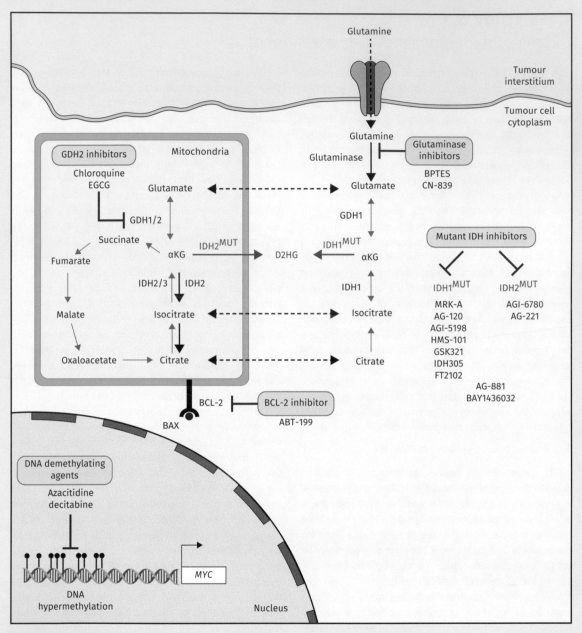

Adapted from Figure 1 in Waitkus, M. S., et al. (2018). Cancer Cell, 34: 1–10.

Figure CS5.1A shows the changes that a mutant IDH1 makes to cell metabolism and how these changes might be useful as therapeutic targets. Mutant IDH may also be targeted directly using small molecule inhibitors such as IDH305. The mtIDH associated hypermethylation phenotype CIMP could be targeted by DNA demethylating agents such as azacitidine.

❯ The hypermethylation phenotype, CIMP, is described in Chapter 4.

a synergistic combination can totally decimate the tumour cell population it may act as a double-edged sword since non-synergistic combinations are more effective than synergistic in delaying evolution of resistance. Combination regimes are often given acronyms based on the drugs used, for example, MOPP (Mustargen, Oncovin, Procarbazine, Prednizone) or CHOP (Cyclophosphamide, Hydroxydaunorubicin, Oncovin, Prednisone).

Treatment that inhibits cancer also inhibits normal cellular proliferation, damages DNA, and prevents normal healing processes, causing side effects. Breast cancer treatments, for example, comes with an increased risk of cardiovascular disease. Cardiac specific mortality is increased by 27 per cent in patients treated with surgery plus radiotherapy in contrast to patients with surgery alone. Anti-angiogenic drugs increase bleeding and prevent wound healing, and often lead to complications such as infection. Most treatments can leave the patient with minimal residual disease (MRD) that is not detected by available methodologies such as MRI or blood tests. MRD has a higher proportion of cancer stem cells and can result in loss of remission and resistance to therapy. Herceptin for example is effective in treating HER2+ breast cancer at extracranial sites but cannot cross the blood brain barrier to remove metastatic cells that have a tropism for brain tissue. Unfortunately, a large proportion of HER2+ breast cancer patients develop brain metastases. Breast cancer is the second most common malignancy associated with brain metastases and around half of patients with HER2+ advanced breast cancer die from secondary brain cancer.

For many cancer types, men and women differ in terms of susceptibility and prognosis. Gender-biased mutational and epigenetic changes may occur in some tumours suggesting a need for sex-specific therapies. Over half of clinically relevant gene changes show sex-biased signatures. Thyroid, bladder, and kidney cancers, for example, show a strong gender effect. Mutation rate for bladder cancer is more than twice in males as in females whereas in thyroid cancer the reverse is true. A high mutation rate for *EGFR* in female patients with non-small cell lung cancer may contribute to enhanced therapeutic responses among women. More research is needed, but this does highlight again that some form of personalized cancer treatment is the way forward.

Chromothripsis is the name given to a large number of chromosomal rearrangements generated in a single catastrophic event. Ionizing radiation or telomere attrition could cause this. In fact, most examples of chromothripsis involve telomere regions. Very rarely, a cell might survive such an event and piece the bits of damaged chromosome back together to create clones with a genetic landscape that confers a significant selective advantage in tumour progression. It is estimated that such an event occurs in 2–3 per cent of all cancers, across many subtypes but particularly in bone cancers. Targeting tumours with short telomeres using DNA damaging agents and telomere repair inhibitors might be counter-productive if it increases the chances of chromothripsis. See Tables 5.4 and 5.5 for a summary of therapeutic problems.

Table 5.5 Problems with current standard therapies

1. Side effects, e.g. Herceptin carries a risk of heart disease.

2. Even extreme doses of combination chemotherapy are unable to cure major killers such as advanced breast cancer.

3. Many women with early breast cancer do no better with adjuvant therapy, deriving from it only toxicity and potential harm.

4. Cost of adjuvant therapy per life saved may not be economically viable.

5. Cost of drugs that are required continuously rather than a short course, e.g. Herceptin and Glivec may be prohibitive.

6. Possibility that no added benefit is gained from adjuvant therapy given to patients with early disease, who do well anyway on conventional treatment.

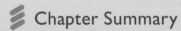

Chapter Summary

- Surgery, radiation, and chemotherapy remain the mainstream of oncological treatments.
- Improved understanding of the molecular basis of cancer has stimulated the development of targeted therapies, including monoclonal antibodies such as Herceptin and small molecule tyrosine kinase inhibitors.
- Immunotherapy, which aims to direct a person's immune system to fight their cancer, is being used to treat cancers such as metastatic melanoma.
- When it comes to cancer treatment, it is becoming evident that one size does not fit all. This has led to a paradigm change in cancer treatment for the future towards a tailored or personalized approach.
- Targeted therapies rely on accurate biomarkers to identify suitable patients and the production of cost-effective drugs.

Further Reading

Baskar, R., Lee, K. A., Yeo, R., and Yeoh, K-W. (2012). 'Cancer and radiation therapy: current advances and future directions'. International Journal of Medical Sciences 9(3): 193–9.

This review explains the principles behind radiation therapy and the new technical advances that keep it at the cutting edge of therapy.

Garrett, M., Sperry, J., Braas, D., Yan, W., Le, T. M., Mottahedeh, J., et al (2018). 'Metabolic characterization of isocitrate dehydrogenase (IDH) mutant and IDH wildtype gliomaspheres uncovers cell type-specific vulnerabilities'. Cancer & Metabolism 6(4): 1–15.

A multi-centre study on patient derived tumour tissue that describes the metabolic differences and pathways differentially enriched between wtIDH and mtIDH tumours.

Jiao, Q., Bi, L., Ren, Y., Song, S., Wang, Q., and Wang, Y. S. (2018). 'Advances in studies of tyrosine kinase inhibitors and their acquired resistance'. Molecular Cancer 17(1): 36. doi: 10.1186/s12943-018-0801-5.

This review explains the role of dysfunctional signalling from tyrosine kinases, how these effects are being targeted and some of the challenges involved.

June, C. H., O'Conner, R. S., Kawalekar, O. U., Ghassemi, S., and Milone, M. C. (2018). 'CAR T cell immunotherapy for human cancer'. Science 359: 1361–5.

This review discusses the opportunities and challenges facing the implementation of CAR T cell therapy for mainstream use in treating cancer.

Ribas, A. and Wolchok, J. D. (2018). 'Cancer immunotherapy using checkpoint blockade'. Science 359: 1350–5.

This paper provides an overview of the mechanisms by which the immune checkpoint proteins CTLA-4 and PD-1 suppress T cell activity and how these checkpoints can be blocked to restore T cell function in tumour cells.

 ## Discussion Questions

5.1 Describe the pros and cons of targeting proliferating cells as cancer therapy.

Hint: think about the effect on cells that are not dividing rapidly such as those in the tumour microenvironment. Describe why there can be a high degree of side effects associated with this strategy.

5.2 How has understanding the IDH mutational status in gliomas facilitated treatment of patients?

Hint: think of the different biological behaviour conferred by IDH. Discuss the changes to metabolic pathways conferred by the IDH status and the possibility of oncogene addiction.

5.3 How could the therapeutic success of immune checkpoint blockade and CAR T cell therapy be improved in human cancers?

Hint: think of toxicities, resistance, and in the case of CAR T cells applying this approach to solid tumours.

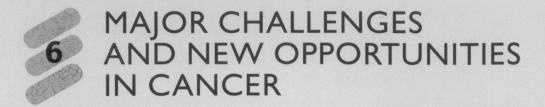

6 MAJOR CHALLENGES AND NEW OPPORTUNITIES IN CANCER

Learning Objectives

- describe the invasion-metastatic cascade and some of the key drivers of this process;
- summarize the treatment approaches that are being explored to prevent metastatic spread;
- describe the mechanisms by which drug resistance occurs including the role of cancer stem cells;
- illustrate with examples, biomarkers that are currently in use or being investigated to personalize the clinical management of cancer patients and the challenges associated with these;
- highlight new technologies that are driving our understanding of tumour heterogeneity and those that are enabling the development of new cancer treatments; and
- describe the developmental status of treatment modalities such as therapeutic cancer vaccines and oncolytic viral therapy.

———————————————

As we have seen from previous chapters, our understanding of signalling pathways underpinning cancer has advanced substantially. In turn, these have informed the development of more targeted therapies and of new treatment modalities, such as immunotherapy. Despite these advances, a number of areas continue to challenge the successful treatment of cancer. This chapter focuses on some of these challenges, namely metastatic tumour spread and resistance to therapy. The cancer genomics revolution combined with the development of new technologies such as single-cell analysis have also created substantial opportunities; leading to the identification and use of biomarkers in the clinical management of cancer and enabling the monitoring of disease resistance. These in turn are facilitating the use of more personalized treatment regimes based on the molecular profile of the patient's tumour. Some of these new directions are also discussed in this chapter.

6.1 Metastasis

Metastatic disease remains largely incurable by current therapeutic approaches and causes the overwhelming majority of cancer-associated deaths—about 90 per cent. Metastasis therefore represents a major challenge to the successful treatment of cancer and many of the mechanisms underpinning metastatic spread remain poorly understood.

Metastasis is a complex process in which primary cancer cells are disseminated from their location to establish secondary tumours at distant sites. It starts with the loss of adhesion of epithelial cancer cells and acquisition of migratory properties that enable cancer cells to invade into the surrounding (local) tissue. This is followed by intravasation of cells into the blood or lymphatic circulation for transit. Cells that survive in circulation extravasate into the parenchyma of a distant target organ. Here the cells may enter a long-lasting dormant state as individual disseminated tumour cells (DTCs) or as multicellular micrometastasis which eventually start to proliferate to form macro-metastases. This process is shown in Figure 2.2.

❯ See introduction to metastasis in Chapter 2.

Epithelial–mesenchymal transition

A critical step in the invasion-metastasis cascade is epithelial–mesenchymal transition or EMT in which carcinoma cells lose many of their epithelial characteristics and take on those more typical of mesenchymal cells. Epithelial cells are stationary, held in position by cell-to-cell interactions and interactions between the cell and the extracellular matrix (ECM) through adhesion molecules like E-cadherin and integrin receptors, respectively. These serve the dual purpose of anchorage and enable signalling between neighbouring cells to take place. Upon activation of EMT, cells lose their neighbouring contacts through the loss of E-cadherin and other markers associated with an epithelial phenotype (for example, cyto-keratin). Instead they begin to express mesenchymal markers such as N-cadherin, vimentin, and fibronectin. This changes epithelial cell morphology from a cuboidal shape with apical-basal polarity to a non-polarized spindle-shaped cell resembling a fibroblast, as shown in Figure 6.1. Transition to a mesenchymal phenotype confers on cancer cells properties critical for metastasis including motility, ability to degrade components of the ECM, and migrate to distant organs.

EMT and the tumour microenvironment

EMT can be induced by multiple extracellular signals received from cells of the tumour microenvironment that converge in the activation of a number of EMT-activating transcription factors (EMT-TFs), mainly of the SNAIL (SNAIL and SLUG), TWIST (TWIST 1 to 3), and ZEB (1 and 2) gene families. These TFs induce the expression of mesenchymal markers and repress epithelial genes, and are activated in different combinations in different cancer types. One of the key signalling molecules implicated in driving EMT is TGF-β, expressed at high levels by tumour cells. The TGF-β/SMAD signalling pathway leads to the expression of Snail, Slug, Zeb-2, and Twist, which in turn results in the repression of CDH1 (encoding E-cadherin) expression and dissolution of cell–cell interactions.

Other key physiological signals and components of the tumour microenvironment involved in determining EMT activation are hypoxia, fibroblasts, inflammation, and macrophages. Figure 6.2 provides an overview of their action.

Figure 6.1 Epithelial and mesenchymal cell morphology. Epithelial cells are cuboidal in shape and stationary held in position through cell-to-cell interactions and cell to ECM interactions. Mesenchymal cells are spindle-shaped, and lose interactions with adjacent cells and with the ECM to become motile.

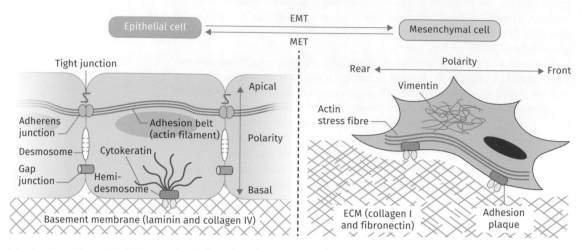

Adapted from Figure 1 in Shibue and Weinberg. (2017). Nature Reviews Clinical Oncology, 14: 611–29.

Hypoxia—Hif-1α signalling in response to hypoxia can induce vascularization to enable tumour growth. Hypoxia-induced Hif-1α can also induce EMT and acquisition of a mesenchymal phenotype through the activation of EMT-TFs. Mesenchymal-like cells with cancer stem cell properties have been shown to trans-differentiate into endothelial cells and thus contribute in part to the tumour endothelium. This process, known as vascular mimicry, is well-studied in glioblastoma and can promote the creation of a vascular system, thus supporting tumour survival and metastasis. Differentiation of cancer cells into endothelial cells is mediated by three signalling pathways involving angiogenesis (VEGF), stem cell differentiation (Notch), and hypoxia (Hif-1α); these are described in Chapters 4 and 5.

❯ See more about hypoxia and VEGF in Chapters 4 and 5.

Cancer-associated fibroblasts—normal fibroblasts residing in the tumour microenvironment can be converted to cancer associated fibroblasts (CAFs) by cancer cells secreting TGF-β. CAFs in turn go onto activate EMT in neighbouring cancer cells through the production of various signalling molecules including TGF-β as well as the matrix metalloproteinases (MMPs). MMPs participate in the cleavage of E-cadherin dissolving cell-to-cell contacts. They also foster local invasion by degrading the basement membrane, allowing cells to migrate into the dense stroma.

Inflammation and tumour associated macrophages—TAMs are a major component of the tumour microenvironment and their increased numbers in the tumour stroma is associated with a higher risk of distant metastasis and poor overall survival in several cancers. TAMs produce inflammatory cytokines such as TNF-α, IL-8, IL-6, and CCL18 that drive tumour development by suppressing T cell immune responses and activate EMT in neighbouring cancer cells through induction of EMT-TFs. TAMs, like CAFs, can also contribute to the activation of EMT through secretion of TGF-β. In addition, TAMs promote local invasion by producing MMPs, cathepsins, and serine proteases to degrade the matrix and migrate through the stroma and into the circulation. Their importance in

Figure 6.2 Components of the tumour microenvironment that promote epithelial–mesenchymal transition. Cells of the tumour stroma (CAFs and TAMs) can secrete soluble factors such as TGF-β, IL-6, and TNF-α, which act on neighbouring cancer cells to activate intracellular signalling cascades, inducing the expression of EMT-TFs. Physiological signals such as hypoxia can also activate EMT-TFs through Hif-1 signalling. These TFs inhibit the expression of epithelial cell markers and promote the expression of markers associated with the mesenchymal phenotype.

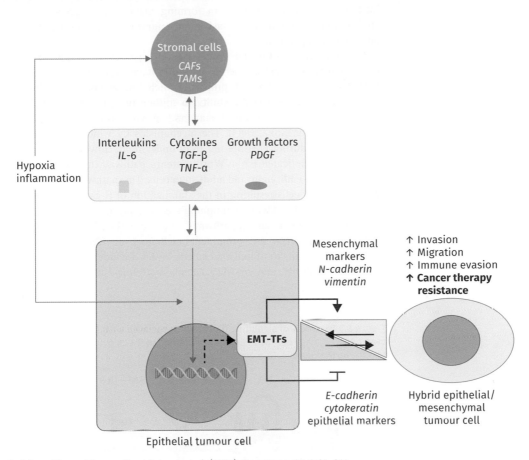

Adapted from Figure 1 in van Staalduinen et al. (2018). Oncogene, 37: 6195–211.

tumour progression is not limited to the initial stages of the metastatic cascade but also extends to the propagation of the tumour at secondary sites as described in 'Metastatic colonization'.

Partial-EMT and tumour seeding ability

Very rarely does activation of the EMT programme in cancer cells lead to its complete transition from an epithelial to a mesenchymal state. Instead EMT activation is often partial, in which the same cancer cell exhibits both epithelial and mesenchymal markers. Thus, EMT programmes are increasingly viewed as generating cells of intermediate phenotypes that reside along an epithelial–mesenchymal spectrum with the possibility of back and forth transitions between the two states. This cell plasticity is important in that it enables the cancer cell to migrate out of the primary tumour site; but it also allows the cancer cell to revert to an epithelial state at the secondary site. The latter process, termed mesenchymal–epithelial transition or MET appears to be an important step in the development of macro-metastatic outgrowth. Cells that undergo complete EMT resulting in fully stable mesenchymal cells are unable to revert to an epithelial state and are inefficient at seeding tumours at secondary sites. In contrast, cells that retain a partial epithelial phenotype are able to undergo MET and are more successful in forming metastatic outgrowth. Thus EMT appears to convey not only invasive and migratory properties but also tumour propagating capability.

As we know from Chapter 1, this ability to perpetuate a tumour is a property of cancer stem cells. A series of studies have shown that cells undergoing EMT display cancer stem cell properties, including acquisition of cancer stem cell surface proteins and the ability to initiate tumours *in vivo*. Cells undergoing EMT can co-express EMT markers (e.g. vimentin, N-cadherin) and stem cell markers (e.g. Sox2, Oct4). The mechanisms by which cells acquire stem cell properties during EMT are not fully understood. However, one key signalling cascade implicated is the Wnt/β-catenin pathway involved in regulating pluripotency in embryonic and adult stem cells. This pathway is aberrantly activated in cancer cells resulting in the stabilization of β-catenin and expression of proteins associated with maintaining stem cell properties. Figure 6.3 outlines some of the major signalling pathways involved in the EMT programme.

❯ See more about cancer stem cells in Chapter 1.

Non-coding RNAs are prominent mediators of EMT cell plasticity, in particular the miR-200 family members, which include miR-200a, miR-200b, miR-200c, miR-141, and miR-429. Members of miR-200 family are associated

Figure 6.3 Overview of signalling pathways associated with EMT and cancer cell stemness. Four key pathways are shown leading to the activation of EMT-TFs. See text for further details. GF = growth factor.

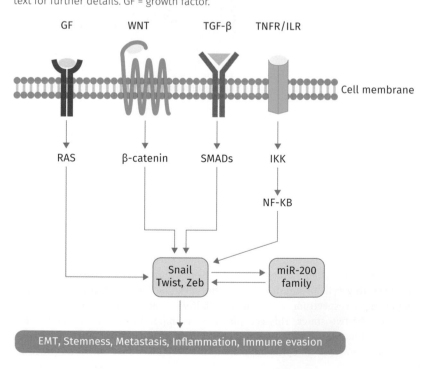

with epithelial cell morphology and their expression is decreased upon induction of EMT by the transcription factors Zeb 1 and 2 binding to miR-200 promoter regions and inhibiting their expression. During MET, the levels of miR-200 family members rise, repressing Zeb 1 and 2 expression and so promoting MET. Hence this double-negative feedback loop acts to regulate the EMT-MET process through a balanced expression of the miR-200 family members and Zeb 1 and 2 EMT-TFs.

Long non-coding RNAs have also been implicated in EMT cell plasticity, although compared to the miRNAs, less is known about their roles. Metastasis-associated lung adenocarcinoma transcript 1 (MALAT-1) is one of the most well-studied lncRNAs. Its ability to induce EMT has been linked to increased levels of Zeb 1 and 2 and reduced E-cadherin expression. Silencing of MALAT-1 reduces Zeb 1 and 2 levels, increases E-cadherin expression, and correspondingly causes MET in some cancer cell types.

Invasion

Cancer cells invade into the surrounding tissue and enter the circulatory system either as individual cells or more typically as groups of cohesive cells. In the latter case cells of the invasive fronts—the leaders—display a more mesenchymal phenotype including reduced E-cadherin whilst the followers retain a more epithelial phenotype and remain attached to each other via adherens junctions (E-cadherin). Invasion of cells into the surrounding tissue requires degradation of the basement membrane and cells at the invasive front express proteinases like the MMPs. One of these is MMP3 activated when its repression by TIMP3 (tissue inhibitor of metalloproteinase 3) is relieved by the action of TGF-β.

Once the cells have invaded through the basement membrane, they can individually or as multi-cellular clumps enter the blood or lymphatic system. Of these circulating tumour cells (CTCs), only an estimated 0.01 per cent are able to form secondary tumours, since the vast majority of cells are unable to survive the physical shearing forces whilst in circulation or escape attack by the immune system. However, CTCs have evolved mechanisms to overcome this challenge through for example, associating with platelets whilst in the blood stream. This interaction can protect the cells from immune attack but may also maintain the EMT programme through the secretion of TGF-β by platelets. Platelets can also secrete ATP, which can act on the P2Y2 receptors expressed on endothelial cells of the blood vessels to make them more permeable. This in turn may support the movement of the CTCs out of the bloodstream and into the parenchyma of distant tissues.

Metastatic colonization

CTCs entering into secondary or distant organ sites may enter a dormant state for many years before a tumour outgrowth is detected. The ability of a tumour to colonize at the secondary site is critically dependent on the local microenvironment CTCs encounter. Primary tumours can prepare a supportive microenvironment at distant organs before the arrival of the cancer cells through the formation of a pre-metastatic niche. Thus the 'seed' the pro-metastatic tumour cells will colonize in the 'soil' where the microenvironment is favourable for metastases as proposed by Paget in his 'seed and soil' hypothesis in 1889. Certain types of cancers metastasize to specific organs—for example, breast cancer commonly metastasize to the bone, brain, and lungs. Mechanisms underpinning this predisposition for disseminated tumour cells to arise within specific organs remain very poorly understood. However, some insights are presented below.

The pre-metastatic niche is established through signalling factors secreted by primary tumours and cells of the tumour stroma that induce the mobilization and recruitment of several cell populations to the secondary organ site. If we take the example of breast cancer tumours metastasizing to the bone, breast cancer cells produce a number of molecules including LOX, parathyroid related hormone protein (PTHrP), IL-6 and MMPs. LOX is induced in response to the hypoxic environment of the tumour and is an enzyme that modifies the extracellular matrix of the bone. This reshaping is an essential step in providing a platform for adhesion of cells subsequently recruited to the site. Secretion of PTHrP, IL-6, and MMPs activate bone-forming osteoblasts to release high levels of the ligand RANKL (receptor activator of NF-κB ligand). RANKL in turn activates osteoclasts to degrade bone. As bone degradation occurs, growth factors are liberated from the bone matrix which go onto stimulate tumour cells to secrete even more factors that then go on to enhance osteoclast activity, thus creating a 'vicious cycle' of cancer growth. Similarly, secretion of chemokines and cytokines by primary cancer cells can recruit immunosuppressive cells, including TAMs and Tregs to the secondary sites. Collectively these molecules work together to create a suitable niche to enable disseminated cancer cells to colonize the distant organ.

More recently, tumour-derived exosomes, small vesicles that carry mRNA, miRNA, DNA or protein, have also been shown to facilitate pre-metastatic niche formation. Breast cancer-derived exosomes carrying miR-105 can increase vascular permeability at secondary organs, thus facilitating initial extravasation and subsequent tumour growth. Exosomes can also express specific combinations of integrin heterodimers on their surface promoting their homing to specific body organs.

 Key Points

- Signals derived from multiple components of the TME and physiological conditions act on cancer cells to drive the invasion-metastatic cascade.
- Cancer cells show a partial-EMT phenotype, a property that is important for their ability to seed metastatic colonies at distant sites.
- Prior to metastatic spread, the primary tumour secretes soluble factors and exosomes, which selectively act on distant organs to prepare a pre-metastatic niche that will favour its colonization.

6.2 Treatment of metastasis

As the pathways underpinning metastasis are not very well understood, identification of molecular targets to which drugs could be directed to treat or prevent metastatic disease lag far behind those for the primary tumour. Indeed, there is huge variability between different tumour types and between patients with the same tumour type in terms of the probability, location, and timing of metastatic spread. This heterogeneity coupled with the plasticity and resistance to treatment of cancer cells presents a serious challenge to the formulation of therapies to treat metastatic disease. Consequently, therapeutic strategies for eliminating metastasis are essentially the same as those directed at the corresponding primary tumour, with the exception of surgery. One of the very few drugs that have been developed to target a metastatic pathway is denosumab. This monoclonal antibody is used in patients with bone metastasis and functions by binding to RANKL reducing osteoclast activity and thus bone metastasis. Alternative approaches to treating metastasis are also being trialled and include targeting the EMT pathway and targeting components of the tumour microenvironment.

Targeting the EMT pathway

One way in which metastatic spread could be prevented is to block signalling pathways that induce EMT. TGF-β signalling is one of the best-characterized EMT-activating pathways, and inhibitors of TGF-β are currently undergoing critical trials. However, this growth factor plays a dual role in carcinogenesis. It can promote EMT and therefore promote cancer progression but it also inhibits cancer formation at early stages of development. Treatment with TGF-β inhibitors would therefore require identification of patients in which it plays a growth-promoting role prior to administration.

An alternative approach is to reverse the process of EMT by directing the transition of mesenchymal cells back to a more epithelial phenotype. One method would be to prevent the activation of EMT-TFs by the delivery of miRNAs involved in their regulation. A number of studies have shown that delivery of miR-200 members in different cancer models including renal, ovarian, breast, and lung cancers can inhibit EMT, reducing primary tumour growth and distant metastasis.

> See more about miRNAs in Chapter 4.

Activation of EMT is associated with drug resistance. Evidence comes from studies that show mesenchymal cells that have undergone EMT are more resistant to drugs than their epithelial counterparts. For example, the small molecule tyrosine kinase inhibitor erlotinib, which targets EGFR to prevent cell proliferation, is less effective in mesenchymal-like cancer cells. Resistance is achieved through increased expression of an alternative tyrosine kinase receptor AXL. Cell signalling occurs through AXL, hence overriding the effects of EGFR inhibition and promoting survival. Inhibitors targeting AXL have been designed and are in clinical trials.

Another mechanism associated with increased resistance in cells that have undergone EMT is through aberrant activation of the Notch and Wnt/β-catenin signalling pathways. As these are associated with maintaining stem cell properties, targeting components of these pathways could be an effective strategy for inhibiting metastasis and overcoming drug resistance associated with cancer stem cells. Drug targets for both signalling pathways are currently in development. For example, PRI-274 is a small molecule inhibitor that blocks the interaction between β-catenin and its co-activator CREB-binding protein (CPB), thus blocking transcription of genes involved in stem cell renewal. Monoclonal antibodies targeting the interaction between the ligand Wnt and its corresponding Frizzled receptor have also been developed. Both these drugs have entered early-phase clinical trials either as monotherapy or as combination therapy for a range of solid tumours.

> See more about targeting cancer stem cell pathways for therapy in Chapter 5.

Targeting the tumour microenvironment

As multiple components of the TME and physiological conditions contribute to the invasion-metastatic cascade, selective targeting of these could aid in preventing or limiting metastatic spread. Current approaches being used or being studied include antiangiogenic therapy and immune checkpoint blockade, both of which are covered in Chapter 5.

> See more about anti-angiogenic therapy and immunotherapy in Chapter 5.

TAMs can also be targeted, and several strategies are being trialled. One of these is to block the production by tumour cells of the growth factor, colony-stimulating factor-1 (CSF-1). CSF-1 stimulates macrophage accumulation within the tumour stroma, and so inhibition of its receptor by a small molecule inhibitor or antisense RNA can reduce macrophage recruitment, and thereby inhibition of cancer cell proliferation and metastasis. Another approach could be to switch M2 macrophages that are tumour promoting to the M1 tumour destructive phenotype. This has been achieved in breast cancer models in which an antibody directed against the IL-10 receptor preventing the pro-tumourigenic IL-10 cytokine from binding, switched M2 macrophages to M1 and triggered an immune response leading to a reduction in tumour mass.

6.3 Molecular markers as the basis of precision cancer medicine

Molecular targeted treatments described in Chapter 5 represent the beginnings of precision cancer medicine which aims to provide 'the right drug, for the right patient at the right time'. Fundamental to this approach is the identification of biomarkers.

A tumour biomarker maybe a molecular change such as DNA, RNA, protein, or metabolite or it may be a process change such as an alteration in tissue morphology or neovascularization. They can be categorized broadly into two types based on their usage. Prognostic biomarkers are used to inform about likely cancer outcomes such as disease recurrence or disease progression in the future. In contrast, predictive biomarkers are used to assess whether an individual will respond favourably or not to a particular treatment.

❯ See introduction to biomarkers in Chapter 2.

High throughput genomic, transcriptomic, proteomic, and metabolomic analysis of tumours have led to the discovery of an abundance of new biomarkers. To be approved for use, discovered biomarkers go through a rigorous development process requiring extensive validation. However, the vast majority fail to translate into the clinic due to not meeting the regulatory requirements of study design, statistical analysis, or assay development method. Below we present examples of predictive and prognostic biomarkers, some approved and some under investigation, to illustrate the utility of biomarkers in personalizing cancer therapy.

Prognostic biomarkers

An early example of an approved prognostic biomarker is prostate specific antigen (PSA). This protein is elevated in patients with prostate cancer—above the upper limit of normal serum concentration of 4.0 ng/ml—and is used in conjunction with more traditional prognostic factors such as tumour grade, tumour size, and lymph node status (positive or negative) to guide prognosis. However, elevated levels of PSA are also observed in benign conditions such as benign prostatic hyperplasia and some men may harbour aggressive forms of prostate cancer despite initial low levels of PSA. Thus more specific biomarkers have been identified, many of which rely on gene (DNA) mutations or changes in gene expression (mRNA or miRNA). Furthermore, instead of measuring a single marker, recent developments have focused on measuring multiple markers—so-called biomarker signatures—to improve discriminatory power. In addition to providing prognostic information, some of these signatures can also predict response to therapy. One of the most well-known examples is the multi-gene panel MammaPrint used in guiding treatment decisions in breast cancer patients.

MammaPrint is able to distinguish early stage breast cancer patients who are at significant risk of metastasis versus those at low risk. The test uses breast cancer tissue samples to measure by microarray, the expression of seventy genes that are implicated in the six original hallmarks of cancer (see Chapter 2). Breast cancer patients are stratified into two groups, low or high risk for disease recurrence, based on the expression levels of these genes. The test also helps identify patients that would benefit from chemotherapy versus those that would not. Studies have shown that results from the MammaPrint test have reduced the administration of adjuvant chemotherapy to patients by 20–30 per cent, without compromising long-term clinical outcomes. Other prognostic tests that work in a similar way, relying on multi-gene expression panels include Oncotype DX (21-gene panel) used with breast cancer patients and myPlan® Lung Cancer (46-gene panel) used with NSLC patients.

The search for prognostic markers is continuing to evolve and more current work has combined interrogation of coding-regions of the genome (mRNA) with non-coding regions of the genome (ncRNA) and with the epigenome (methylation status). For example, a recent study showed that a signature comprised of protein-coding genes (*XPO1*, *BRCA1*, *HIF1*-α, and *DLC1*), non-coding miRNA (miR-21) and promoter methylation of the developmental gene *HOXA9* could identify stage 1 lung cancer patients who were at high risk of metastasis and would benefit from adjuvant chemotherapy. This statistical association was stronger when all three marker classes were combined than when utilizing coding or non-coding expression data alone. Such a multi-omics approach allows for a more comprehensive assessment of the tumour landscape to categorize patient risk and inform treatment, but are at very preliminary stages of development.

Discovery of biomarker signatures based on protein-expression have lagged behind those based on genetic or transcriptional profiling due to the technological difficulties associated with interpreting complex proteomes. However, one approved for use is the OVA2 test for ovarian cancer. This utilizes seven serum-based protein biomarkers: CA125 (a cancer antigen elevated in some tumours including ovarian) combined with transthyretin, apolipoprotein A-1, transferrin, and β2-microglobulin, and more recently, follicle stimulating hormone and human epididymis protein 4. The test is used to identify whether a tumour mass detected through imaging is malignant or benign, prior to surgical removal of the mass. Patients with malignant tumours can be referred to an oncologist for appropriate treatment, improving overall outcomes.

Predictive biomarkers

Breast cancer has led the way in the use of predictive biomarkers; prime examples are the use of ER status to predict response to endocrine therapy and HER2 status to predict response to the monoclonal antibody trastuzumab. The use of predictive biomarkers has expanded to other cancers including colon, lung, melanoma, bladder, cervical, and glioblastomas of which some are described below.

The *K-RAS* genes are amongst the most frequently mutated in colorectal cancer. Somatic mutations within *K-RAS* are found in 30–40 per cent of cases

❭ See more about trastuzumab therapy in Chapter 5.

with the majority of mutations occurring in codons 12 and some in codons 13 and 61. Patients positive for *K-RAS* mutations are resistant to anti-EGFR monoclonal antibodies cetuximab and panitumumab, directed against the extracellular domain of the EGFR. Thus colorectal cancer patients are genotyped for *K-RAS* mutations and anti-EGFR therapy administered only to those with the wild-type *K-RAS* gene.

In addition to gene mutations, miRNAs can also serve as predictive biomarkers. One area in which serum miRNAs are currently being explored is in HER2+ breast cancer patients. Although HER2 overexpression is a prerequisite for effective response to anti-HER2 therapy, a fraction of HER2-positive patients fail to respond to trastuzumab. A number of miRNAs are associated with trastuzumab response; for example, miR-21 overexpression in primary breast cancer is associated with a poor response to trastuzumab therapy. Similarly, a recent study has shown that a signature of four miRNAs—miR-940, miR-451a, miR-16-5p, and miR-17-3p—can predict sensitivity to trastuzumab in metastatic patients. High levels of miR-940 and reduced levels of miR-451a, miR-16-5p, and miR-17-3p are associated with resistance to treatment with trastuzumab whilst the converse expression profile is observed in patients who respond. miR-940 is released primarily from tumour cells whilst miR-451a, miR-16-5p, and miR-17-3p are released from immune cells, mainly T lymphocytes and monocytes. These miRNAs induce resistance by modulating the expression of the tumour suppressor *PTEN* (miR-940) and of the pro-oncogenic *IGF1R* and *SRC* (miR-451a/miR-16-5p and miR-17-3p, respectively). Increased levels of tumour infiltrating lymphocytes are associated with trastuzumab benefit in early and advanced breast cancer patients and this benefit could well be mediated through the action of secreted miRNAs which downregulate the expression of key oncogenic targets within neighbouring tumour cells.

Tumour heterogeneity: a challenge to precision treatment

A challenge to the use of biomarkers in predicting overall cancer survival and guiding treatment is intra-tumour heterogeneity. As we know from Chapter 1, tumours are comprised of sub-populations of cells with distinct genotypic and epigenetic profiles resulting in substantial variation in cell morphology, metabolic requirements, and in metastatic potential.

❯ See more about tumour heterogeneity in Chapter 1.

Molecular biomarker testing has traditionally relied on the analysis of samples extracted from a single needle biopsy or surgical resection sample. These methods capture the molecular and/or histological profile of some tumour cells but not of others. For example, bulk sequencing of DNA extracted from tumour biopsies provide an average picture of all the tumour clones within the sample but does not identify the molecular differences that exist between individual tumour cells. Thus therapy selected to target a particular mutation may eradicate a population of tumour cells, allowing those that do not carry the mutation to survive and cause tumour recurrence.

The molecular profile of tumour cells also changes during the course of disease progression and thus treatment based on markers identified in the primary tumour may not accurately reflect the molecular make-up of the corresponding metastases. Furthermore, treatment with chemotherapy or targeted drugs can alter the molecular landscape of the tumour, enriching for drug-resistant cells. Better characterization of spatial heterogeneity can be achieved by sampling multiple regions of the primary or metastatic tumour.

Similarly temporal heterogeneity can be captured by longitudinal monitoring of molecular changes through obtaining repeated biopsies of tumour cells. However, biopsies are invasive, costly, and in the case of metastases rarely performed. Development of recent techniques that allow singe cell analysis is overcoming the need for repeated biopsies, and enabling a more accurate understanding of tumour heterogeneity. Two such methods are illustrated In Scientific approach panel 6.1.

Both methods highlight that the mutational profile of the cancer evolves in response to treatment. The methods identify cell populations that respond to therapy, those that are intrinsically resistant, thus failing to respond to initial treatment, and others that respond favourably to treatment but then acquire resistance. Resistance can arise through many different mechanism, some of which are summarized in Table 6.1 and shown schematically in Figure 6.4.

Insights from such studies have the potential to improve therapeutic design and lead to better patient outcomes. Based on the expression of specific markers within tumour cell populations, upfront drug combination strategies could be utilized that co-target drug-sensitive populations and pre-existing resistant sub-clones to avoid or delay the emergence of resistance. By monitoring mutational changes that occur during patient treatment emergent drug-resistant cell populations could be identified and targeted more quickly. Additionally, second-line treatments can be applied that more accurately reflect the molecular make-up of the tumour or metastases. However, identification of mutational profiles of each cancer patient before and during treatment using single-cell analysis techniques is not currently feasible to use as a routine clinical approach. Both the technology and the infrastructure would need to be developed to reach this goal.

Scientific approach panel 6.1
Single cell analysis methods for characterization of tumour heterogeneity

One method of single cell analysis is STAR-FISH (Specific-To-Allele PCR–FISH), developed by Janiszewska and colleagues in 2015. The method uses *in-situ* PCR to detect point mutations combined with FISH to detect copy number variation within the same cell. As shown in Figure SA 6.1A, using formalin-fixed, paraffin-embedded breast cancer samples derived from twenty-two HER2+ patients, the authors mapped *PIK3CA* (which encodes the catalytic subunit of PI3K) mutation status (by *in-situ* PCR) and *HER2* amplification (by FISH) in patients before and after treatment with chemotherapy. Anti-HER2 therapy exerts its effect in part through the PI3K/AKT signalling pathway and mutated *PIK3CA* is associated with resistance to trastuzumab treatment. Thus *PIK3CA* was used as a marker to monitor resistance to therapy. Prior to therapy, the majority of tumour cells carried a *HER2* amplification and were wild-type for *PIK3CA*, with very small fractions of cells with mutant *PIK3CA*. However, after neoadjuvant chemotherapy, *PIK3CA* mutant cells were enriched as were cells that lacked the *HER2* amplicon, increasing the risk of resistance to subsequent HER2-targeted therapy (see Figure SA 6.1B).

Another technique that allows monitoring of tumour heterogeneity is longitudinal tracking of the molecular profiles of CTCs or cell-free plasma DNA (cfDNA) extracted from the patient's blood. One of the first studies to

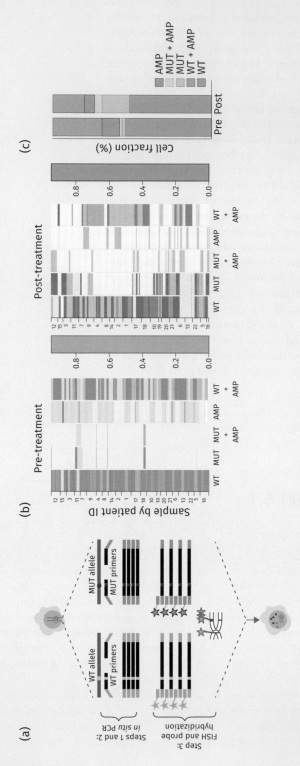

Figure SA 6.1 Single cell analysis using STAR-FISH (Specific-To-Allele PCR—FISH). (a) Schematic outline of the STAR-FISH method performed on a cell heterozygous for a mutation. In steps 1 and 2, *in-situ* PCR using a mixture of primers to the wild-type (WT: green) and mutant (MUT: red) *PIK3C* gene is performed. In step 3, cells are hybridized with a FISH probe mix comprised of fluorescently labelled probes specific for the WT or MUT PCR products, probe to detect HER2 copy number variation (HER2 BAC: magneta) and CEP17 (detects centromere of chromosome 17, the chromosome on which *HER2* is located) (blue). (b) Frequency of each cell type before (pre)and after (post) treatment with chemotherapy. Colour intensity indicates frequency. Based on the results, each cell was assigned to one of five categories: WT—a cell with wild type *PIK3CA* signal and no *HER2* amplification; MUT—a cell with mutant *PIK3CA* signal and no *HER2* amplification; AMP—a cell with no *PIK3CA* signal and *HER2* amplification; MUT+AMP—a cell with mutant *PIK3CA* signal and *HER2* amplification; or WT+AMP—a cell with wild type *PIK3CA* signal and *HER2* amplification. (c) Graph shows the mean percentage of each cell type in all areas and samples combined before (pre) and after (post) therapy. Adapted from Figures 1 and 3 in Janiszewska et al. (2015). Nature Genetics, *47*: 1212–19.

show such liquid biopsies could be used to track genetic changes arising in response to treatment in real-time was Siravenga and colleagues in 2015. These authors carried out next-generation sequencing of cfDNA in colorectal cancer patients following treatment with EGFR directed antibodies cetuximab or panitumumab. They identified the emergence of clones enriched with genes associated with resistance to anti-EGFR therapy including *K-RAS*, *N-RAS*, and *FLT3* mutations and *HER2*, *MET*, and *K-RAS* amplifications in patients unresponsive to, or with acquired resistance to anti-EGFR blockade. Studies such

as these are revealing hugely important insights into how cancer progresses and evolves in response to treatment. Further development of methodologies that make it possible to measure accurately multiple parameters (genome, epigenome, transcriptome, proteome) in single cells will add to our understanding of tumour heterogeneity and enable a greater degree of precision in cancer therapy.

❯ See more about liquid biopsies in chapter 2.

 Key Points

- Identification of biomarkers is fundamental to the development of more precise diagnostic and therapeutic approaches.
- Tumour relapses or metastases are often due to the enrichment of drug-resistant tumour cell populations that survive treatment and repopulate.
- Technologies such as single-cell analysis are enabling better monitoring of the evolution of drug resistance and combined with the use of biomarkers, should facilitate more personalized treatment regimes for cancer patients.

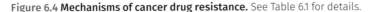

Figure 6.4 Mechanisms of cancer drug resistance. See Table 6.1 for details.

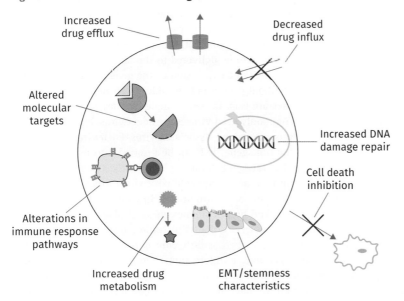

Table 6.1 Mechanisms of drug resistance

Mechanism	Example
Mutation acquired in the protein against which the drug is targeted thus rendering the protein insensitive to the drug.	Acquisition of T790M mutation in the ATP-binding domain of the EGFR in NSCLC. This single amino acid change reduces the accessibility of ATP-competitive TKIs to this domain and so prevents EGFR blockade.
	Third and second generation TKIs have been developed to overcome this resistance.
Mutation within a downstream component of the same signalling pathway to which a drug is targeted.	Acquisition of resistance to trastuzumab in HER2+ cancers through mutations in *PIK3CA*, a signalling molecule that works downstream of *HER2*.
Activation of the same downstream signalling pathway through engaging proteins that act in parallel to the one being targeted by the drug.	Development of resistance through the use of the alternative RTK AXL when small molecule inhibitors are directed against EGFR, as described in 'Epithelial–mesenchymal transition'.
Elevated levels of the ATP binding cassette (ABC) proteins. ABC proteins are membrane receptors that are able to efflux drugs across the plasma membrane.	Overexpression of these proteins have been observed in several cancers and in cancer stem cells and leads to the pumping of chemotherapeutic agents and other drugs out of the cells, reducing their effectiveness.
Inactivation of pathways through which immune checkpoint inhibitors mediate their effects.	Mutations in the *JAK 1* and *JAK 2* genes result in non-functional truncated JAK proteins, which in turn reduces the expression of PDL-1 and hence renders ineffective immune checkpoint inhibitors like pembrolizumab (see Section 5.3).

6.4 Gene therapy with a specific focus on oncolytic viruses

In the broadest sense, the aim of cancer gene therapy is to transfer functional genes to replace faulty genes specifically to cancer cells whilst leaving normal cells unharmed. Although simple in concept, the technical challenges are substantial: the gene construct has to be delivered to the target cell or tissue and, once inside the cell, maintain stable expression of the gene product at the correct dose to be effective, without generating any side effects or an immunogenic response. Typically, viruses are used to deliver genes as they are able to infect human cells naturally. Commonly used viruses are adenovirus, lentivirus, retrovirus, herpes simplex virus, and adeno-associated virus. Each virus class varies in tropism, the size of the transgene that can be inserted into its genome and its capacity to express the transgene stably or transiently. Long-term (stable) expression of months, up to years, is possible with viruses that integrate into the host genome, such as lenti- and retroviruses. In contrast, non-integrating viruses such as adenovirus and herpes simplex virus display transient expression, diminishing within a few weeks of administration. Integrating viruses can pose a significant safety risk, as there is little control over where the virus will insert into the patients genome. Random integration into coding sequences or in promoter regions can lead to altered gene regulation and in some cases tumourigenesis. Whilst efforts have been made to develop safer viral vectors, the risk of insertional mutagenesis is still not completely eliminated.

One of the earliest approaches to cancer gene therapy included delivering a wild-type *TP53* gene to cancer cells that lacked functional p53 using adenovirus as the vector. To disable the virus' ability to replicate within the host, its *E1A* gene is deleted and a functional human p53 gene is added. This therapy, termed Gendicine™, was the first to be approved for use in humans in 2003 by China's Food and Drug Administration for the treatment of head and neck squamous cell carcinoma (HNSCC), based on clinical trials data showing patient improvement. Subsequent studies were unable to replicate these improvements and China currently remains the only country licensed to use Gendicine.

An alternative to using replication-deficient viruses is to use replication-competent viruses. Oncolytic viruses are designed to preferentially replicate in, and lyse cancer cells. This is followed by localized spread of viral progeny released from the lysed cell into neighbouring tumour cells, ultimately leading to a reduction in tumour mass.

To achieve selective replication in cancer cells, viral genes whose products normally supress the action of proteins involved in regulating cell cycle progression are commonly deleted. One of the first oncolytic viruses to be developed was ONYX-015—an adenoviral construct which carries a deletion in its *E1B-55K* gene. This gene encodes a 55 kDa protein that is able to bind to and inhibit p53 activity. When the *E1B-55K* gene is deleted, viral replication is restricted to p53-deficient cells, thus triggering tumour cell lysis. However, studies in which ONYX-015 was tested showed poor correlation between p53 mutation status and response to treatment. Further development of the therapy was halted in 2003 amidst questions about its mechanism of action. Improvements in designing adenoviruses have since continued, focusing on alternative approaches for restricting replication to tumour cells as well as improving the ability of viruses to bind to and enter cancer cells. No adenovirus-based oncolytic viruses have been approved for use although many are in clinical trials. The exception is the ONXY-015-related H101 oncolytic viral therapy, approved for use in China.

After many years of setbacks, the first oncolytic virus therapy was approved in the USA in 2015, followed by Europe and Australia, for use in patients with advanced melanoma. Talimogene laherparepvec (T-VEC), marketed under the tradename Imlygic, is a genetically modified type 1 herpes simplex virus (HSV-1) designed to selectively replicate in tumour cells and induce a host immune response. HSV-1 is a human pathogen that can infect skin and peripheral nerves. It remains latent but during times of stress causes recurrent fever blisters. Its ability to cause infections is removed through deletion of both copies of the gene encoding ICP34.5 (infected cell protein 34.5). Additionally, this deletion provides tumour selective replication. ICP34.5 is able to disrupt protein kinase R (PKR) activity which is involved in blocking viral replication, thus enabling viral propagation. In cancer cells, the PKR pathway is commonly disrupted and so ICP34.5 deletion promotes viral growth in cancer cells but not in healthy cells as shown in Figure 6.5. T-VEC also carries a deletion of the *NS* gene, which prevents ICP47 from blocking antigen presentation. To facilitate a wider anti-tumour response, the gene encoding GM-CSF is also included. Following administration of T-VEC, the virus exerts its action in two main ways, one it causes the cancer cells to lyse and two, the GM-CSF promotes the maturation of dendritic cells, which in turn activate T cells. T cells can then migrate to tumour sites to induce T cell mediated cell death.

T-VEC in advanced melanoma patients as a monotherapy has shown significant improvement in durable and objective response rates. However, immune responses can be evaded through the production of immunosuppressive checkpoint proteins on the surface of T cells and tumour cells, such as CTLA-4 and PD-1. To enhance the anti-tumour activity of T-VEC, studies assessing T-VEC

Figure 6.5 Mechanism of selective replication in cancer cells by the oncolytic virus TVEC. (a) When the ICP34.5 gene is deleted, the virus is prevented from replicating in non-tumour cells through the action of a functional PKR (b) In cancer cells, PKR is mutated and so viral replication occurs leading to cancer cell death. (c) Tumour cell lysis leads to the release of tumour antigens and GM-CSF. The latter promotes maturation of dendritic cells, which in turn activate T cells to mediate tumour cell death. (Note: TVEC also carries an ICP47 deletion, not shown in this figure.)

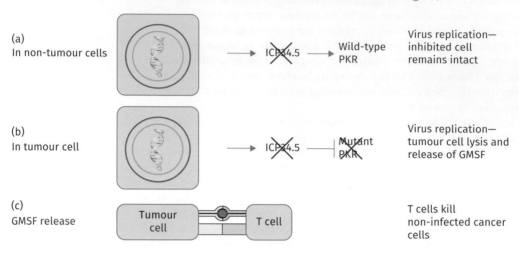

in combination with immune checkpoint inhibitors, such as ipilimumab and pembrolizumab, are in progress for use in melanoma but also for other solid tumours. Oncolytic therapy as a class of cancer therapeutics are likely to expand in use to treat a variety of cancer types using various viral platforms, as research in this area continues to progress.

6.5 New directions in immunotherapy

Recent advances in cancer immunotherapy as described in Chapter 5 have dramatically shifted the landscape of cancer treatments. However, immunotherapy is currently limited to the treatment of metastatic tumours in patients that have exhausted all current treatment regimes. Responses are also variable, with some patients responding whilst others do not. Therefore, an area of current research is to understand the biological mechanisms underpinning these variable responses and to identify biomarkers that could predict which patients are likely to respond to these therapies. Additionally, approaches to generate more personalized modes of immunotherapy are being investigated. Two of these are discussed below: editing T cell genomes and generating personalized therapeutic vaccines.

❯ See more about immunotherapy in Chapter 5.

Ex vivo genome editing of T cells

In *ex vivo* genome editing, cells are sampled from a patient, modified to correct for a defective gene and then the cells with the corrected gene sequence are reintroduced back into the patient. We have come across the *ex vivo* approach with CAR T cell therapy in Chapter 5 in which the extracellular domain of the TCR is replaced by a single chain antibody that recognizes a tumour antigen. Researchers are building on this by precisely editing the genome of T cells using the CRISPR-Cas 9 genome editing technology. Lu and colleagues at Sichuan University in Chengdu removed T cells from a patient's blood, deleted the gene encoding for the checkpoint protein PD-1, expanded the edited cells *in vitro*, and injected them back into the patient who had metastatic NSCLC. This study, conducted in 2016, was the first trial to inject a person with cells that contain genes edited with CRISPR-Cas 9 and involved treating ten patients, each receiving between two to four injections. A phase 1 study in the USA has also started, using the *ex vivo* approach and will enrol up to eighteen patients with one of three cancer types, multiple myeloma, sarcoma, and melanoma. T cells extracted from the patient's blood will be CRISPR gene edited to delete the endogenous PD-1 and TCR genes. Additionally, the cells will be engineered to express a TCR that can bind to the tumour-associated antigen NY-ESO-1. NY-ESO-1 is a cancer testis antigen whose expression is normally restricted to male germline cells but is upregulated across many tumours and so T cells can be directed to bind to the tumour cells via this antigen.

The CRISPR-Cas 9 system comprises of guide RNA (gRNA) sequence complementary to the target sequence combined with the Cas 9 endonuclease. In the *ex vivo* approach, it is delivered to the cell as a ribonucleoprotein (RNP) complex using electroporation. When co-delivery of donor DNA is required as a template to introduce a new sequence into the genome, integrating viruses such as adeno-associated virus and integrase-defective lentiviral vectors are used. Thus the obstacles to using CRISPR-Cas 9 are similar to those associated with gene therapy, described in the section 'Gene therapy with a specific focus on oncolytic viruses', including the possibility of nuclease-associated off-target effects and ensuring safe and efficient delivery of CRISPR-Cas 9 and donor DNA to the target cell. It may be possible in the future to transfer donor DNA into T cells without using viral vectors as shown by work recently published by Roth and colleagues. These authors were able to replace an endogenous TCR with a new TCR designed to target the tumour-associated antigen NY-ESO-1 by co-electroporation of T cells, the CISPR-Cas 9 RNP complex and the DNA template. The engineered T cells were able to recognize the tumour antigen and mount an anti-tumour immune response *in vitro* and *in vivo*. The success of this technique was based on the ratio of T cells, CRISPR-Cas9, DNA, and the electrical-field parameters which were discovered through an effort described as 'Herculean'.

Gene editing is generating much excitement and is likely to usher in a new era in the treatment of cancer. However, it is at very early stages and although a small number of gene editing studies have entered the clinic, major issues need to be tested including the specificity and safety of the CRISPR-Cas 9 system.

> See more about immune checkpoint blockade therapy in Chapter 5.

Therapeutic cancer vaccines

The aim of therapeutic cancer vaccines is to treat established disease by invoking a T cell based immune response against the tumour. Vaccines are composed of tumour antigens and **adjuvants**; the latter are non-antigen specific triggers of the immune system that enhance the immune response against the antigen. A number of different vaccine types have been developed but the vast majority are in the research phase.

The development of cancer vaccines is challenging; one of the key reasons being the identification of antigens that can elicit a strong immune response upon vaccine administration. The vast majority of vaccines to date have utilized tumour-associated antigens. As they are shared across many cancers, the vaccine can be administered to different patients with the same cancer type but also across different types of cancers. There are three broad approaches to cancer vaccine formulations; protein or nucleic acid-based vaccines and whole cell-based vaccines. Each of these is described in turn below.

Protein or nucleic acid-based cancer vaccines

In protein-based cancer vaccines, short peptides (eight to ten amino acids) representing the epitope which is presented to T cells via MHC molecules are used. Multiple peptides comprising different antigens and targeting several T cell clones can also be given at the same time as can longer peptide sequences or full-length proteins. A protein-based cancer vaccine currently being explored in clinical trials is NeuVax. The antigen is a peptide sequence derived from the extracellular domain of the HER2 protein. It is used in conjunction with the adjuvant GM-CSF. Antigens encoded by DNA or mRNA sequences can also be delivered to the tumour cells *in vivo* using a delivery system such as viral vectors. The resulting peptides are presented by MHC molecules on the surface of APC cells to stimulate an immune response. Nucleic acid-based cancer vaccines have shown limited success thus far. However, improvements in design and delivery methods are ongoing with a number of nucleic acid-based vaccines in clinical trials.

Vaccination with whole cells

An alternative strategy to designing protein or nucleic acid-based cancer vaccines is to vaccinate patients with whole tumour cells. The rationale underpinning this approach is that cancer cells harbour numerous tumour antigens and therefore are likely to elicit a stronger T cell response given that they contain multiple epitopes for priming both CD4+ and CD8+ cell function. It also overcomes the requirement to identify the MHC I restricted peptide region. Tumour cells used for vaccination can be autologous, derived from the patient or allogenic, derived from established non-self cancer cell lines. An example of a whole tumour cell-based vaccine is GVAX. It is comprised of tumour cells which have been irradiated to prevent cell division and genetically modified to express GM-CSF. GVAX formulations for a number of cancers including prostate and pancreatic cancers are currently in clinical trials as combinatorial partners with checkpoint inhibitors and/or other drugs.

Vaccine preparations can also comprise of whole dendritic cells as in the case of Sipuleucel-T, currently the only therapeutic cancer vaccine to be licensed for use. In dendritic cell cancer vaccines, immature dendritic cells are isolated from a patient, loaded with antigen in the form of peptides, proteins, or tumour cells either with or following maturation with pro-inflammatory cytokines and infused back into the patient. Sipuleucel-T consists of autologous antigen presenting cells loaded with the tumour associated antigen prostate acid phosphatase and the adjuvant GM-CSF. When infused back into the patient, the dendritic cells present the tumour antigen to T cells inducing their activity. Sipuleucel-T was approved in 2010 for the treatment of metastatic prostate cancer based on modest improvement in overall survival, approximately 4 months, observed in a Phase 3 clinical trial.

Personalized cancer vaccines

The examples of cancer vaccines provided above use tumour associated antigens rather than neoantigens. However, tumour associated antigens can induce autoimmunity against corresponding normal tissues in which the antigen is also expressed and generate a weak T cell response. In contrast, the use of

Scientific approach panel 6.2
Personalized cancer vaccines

Two Phase 1 clinical trials highlight the potential of personalized cancer vaccines in treating cancer. Patrick Ott and colleagues recruited six patients with advanced stage melanoma for their study. Tumour specific mutations were identified by whole exome sequencing of melanoma DNA and normal cell DNA from each patient. Neoantigens with the highest probability of being presented by the patients MHC molecules were selected using computational algorithms that predict MHC–antigen binding affinities. Each participant was vaccinated with synthetic long peptides representing up to twenty neoantigens specific to each patient, together with an adjuvant. The vaccines generated CD4+ and CD8+ responses which targeted fifty-eight (60 per cent) and fifteen (16 per cent), respectively, of the ninety-seven unique neoantigens used across the six patients. Of the six patients, four who had enrolled with stage 3 melanoma showed no sign of tumour recurrence in a follow up period of thirty-two months after vaccination. The remaining two, enrolled with stage IV melanoma showed tumour recurrence shortly after vaccine administration. However, when treated with the PD-1 receptor blocking antibody pembrolizumab, their tumours regressed.

The second study by Ugur Sahin et al. also recruited patients with advanced stage melanoma but used individualized mRNA-based cancer vaccines. These authors used RNA-sequencing to identify the full repertoire of somatic mutations that are expressed using RNA extracted from each patient's tumour sample. From these data, neoantigens with the highest binding affinity to MHC molecules were selected using computational methods. Each patient received a personalized vaccine comprising of up to ten mRNA neoantigens. T cell responses were observed against seventy-five (60 per cent) of the 125 mutations represented in the mRNA vaccines. Of the thirteen melanoma patients recruited, eight were disease-free at the time of vaccine administration and remained so throughout the follow-up period of twelve to twenty-three months. Of the five melanoma patients with metastatic disease at the time of administration, two of these achieved an objective response (at least a 30 per cent decrease in tumour diameter) whilst one achieved a **complete response** in

Figure SA 6.2 Generation of personalized vaccines. DNA or RNA are extracted from tumour cells and from normal tissue cells from the same patient. Somatic mutations are identified through whole exome sequencing (WES) and then those that are expressed are identified by RNA-sequencing. HLA (MHC) typing is carried out using DNA from normal cells. Neoantigens that bind to HLA/MHC molecules with the greatest binding affinity are identified through algorithm prediction models. Candidate epitopes are validated and incorporated into personalized cancer vaccines for the patient.

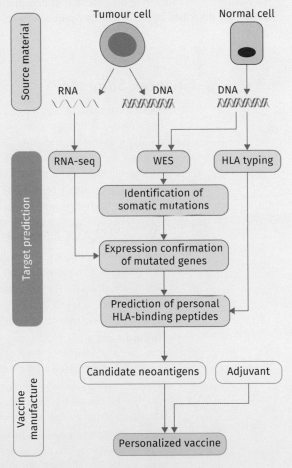

Figure 4A in Hu, Z., et al. (2017). Nature Reviews Immunology, 18: 168–82.

combination with anti-PD-1 therapy. Figure SA 6.2 gives an overview of how these personalized vaccines were generated.

Both these studies show that personalized vaccines are feasible, safe, and can generate T cell immune responses in melanoma. Clinical trials using personalized vaccines are now taking place in multiple tumour types using peptide-based vaccines or RNA-based vaccines in combination with approved antibodies that block immune checkpoint proteins.

personalized vaccines in which neoantigens are derived from individual patients could provide a more effective anti-tumour response. This approach requires identification of somatic mutations carried within the cancer cell genome of each individual patient, and so was not feasible until recently with the advent of next-generation sequencing technologies. Two papers published in the journal *Nature* in 2017 illustrate the potential of personalized cancer vaccines. Both studies are described in the Scientific Approach Panel 6.2.

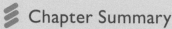

Chapter Summary

- Metastatic disease remains largely incurable by current therapeutic approaches and is responsible for the majority of cancer deaths.
- The ability of cancer cells to transition back and forth between epithelial and mesenchymal states appears to be an important step for a number of cancers in disseminating cancer cells and seeding new tumours at distant locations.
- Many primary tumours establish macrometastases at specific organ sites. This is facilitated through cross-talk between cells of the primary tumour and the distant organ to form a pre-metastatic niche.
- Cells undergoing EMT are associated with drug resistance and this is linked via the EMT programme to the acquisition of the cancer stem cell phenotype.
- Thus one approach to treating EMT and overcoming drug resistance is to develop therapies that target cancer stem cells.
- Treatment with cancer drugs can drive tumour heterogeneity, leading to the evolution of resistance to drugs to which patients where initially responsive.
- Prognostic and predictive biomarkers are increasingly being developed, validated, and used in the clinic to guide a more personalized approach to cancer treatment, based on the molecular profile of the patient's tumour.
- Liquid biopsies are an important tool in monitoring how a patient responds to a particular treatment. This technology, combined with the use of key biomarkers, is likely to provide a new level of precision tumour therapy in the future.
- Immunotherapies continue to be developed to enhance their effectiveness using strategies such as precise editing of T cell genomes and the use of tumour-specific antigens to promote a more effective anti-tumour response.
- Oncolytic viruses as a cancer therapy has had many setbacks but recent approval of T-VEC suggests that with continued improvements in viral design to improve safety and specificity, this treatment modality may expand.

Further Reading

Bilsland, A. E., et al. (2016). 'Virotherapy: cancer gene therapy at last?' F1000Research 5: F1000 Faculty Rev-2105.

This review provides an overview of adenovirus and herpes virus based onco-lytic viral therapies and how they are being developed to improve their efficacy and safety.

Dagogo-Jack, I. and A. T. Shaw (2017). 'Tumour heterogeneity and resistance to cancer therapies'. Nature Reviews Clinical Oncology 15: 81–94.

In this review, the authors describe how improved characterization of spatial and temporal heterogeneity using emerging technologies could enable a better understanding of the process and its associated consequences.

Hu, Z., et al. (2017). 'Towards personalized, tumour-specific, therapeutic vaccines for cancer'. Nature Reviews Immunology 18: 168–82.

This paper provides an overview of how this field is progressing and its future potential, including a summary of the first in-human clinical trials of personal-ized cancer vaccines.

Nair, M., et al. (2018). 'Cancer molecular markers: a guide to cancer detection and management'. Seminars in Cancer Biology, 52(1): 39–55.

This paper includes a detailed summary of biomarkers that have been discov-ered across a range of cancer types.

Nikolaou, M., et al. (2018). 'The challenge of drug resistance in cancer treatment: a current overview'. Clinical & Experimental Metastasis 35(4): 309–18.

This review describes the major mechanisms of acquired drug resistance and strategies to overcome them.

Shibue, T. and Weinberg, R. A. (2017). 'EMT, CSCs, and drug resistance: the mech-anistic link and clinical implications'. Nature Reviews Clinical Oncology 14: 611 –29.

This paper presents an overview of the evidence connecting activation of the EMT programme with the cancer stem cell phenotype and resistance to therapy. It summarizes how this knowledge can be used to improve the treatment of cancer patients.

 ## Discussion Questions

6.1 What are some of the major challenges to the treatment of metastatic tumours? How could these be addressed?

Hint: consider the link between EMT, cancer stem cells, and resistance to therapy.

6.2 Using specific examples, outline how a clinician may use biomarkers to guide the management of cancer patients.

Hint: consider a range of tumour biomarkers (DNA, RNA, protein) to illus-trate your answer.

6.3 Outline some of the technological developments that are paving the way for the development of more personalized cancer treatments.

Hint: think of methods for analysing tumour heterogeneity, techniques used in biomarker discovery and genetic engineering technologies.

GLOSSARY

Adenoma: A benign tumour formed from glandular structures in epithelial tissue

Adjuvant: An immunostimulatory agent designed to enhance the immune response to a vaccine

Adoptive cell therapy: A treatment in which antitumour lymphocytes are identified, expanded *in vitro* and then infused into a patient with cancer. The lymphocytes can be autologous or allogenic

Allogenic: Cells derived from another person (donor)

APBs: ALT-associated promyelocytic (PML) nuclear bodies. Specialized PML bodies comprising of PML and telomeric DNA and proteins

Asymmetrical cell division: Method of cell division employed by stem cells in which one cell is a copy of the parent cell and the second is directed into a differentiation programme

Atypia: Structural abnormality or irregularity in a cell

Autologous: Cells derived from the same person being treated (patient's own)

Benign: A tumour that is non-invasive and therefore not malignant

Biomarker: A molecular change (DNA, RNA, protein, or metabolite), produced by cancer cells that can be used to diagnose, and predict prognosis or response to therapy in a cancer patient

Biopsy: Tissue removed from a living body for diagnostic purposes

Bystander effect: Induction of biological effects in cells that are not directly caused by irradiation (in this context)

Cancer stem cell: Able to self-renew and generate non-stem cell descendents. Able to seed a tumour when transplanted into a host such as an immunocompromised mouse

Cancer syndrome: Gene mutations are carried through the germline that predispose individuals to a higher than normal risk of certain cancer types. Certain patterns of cancer may be seen within families (e.g. developing cancer at an early age, multiple members of the family affected, more than one cancer type in a single individual). See also familial cancer

Carcinogen: A substance capable of causing cancer in an organism

Carcinoma: A cancer arising in the epithelial tissue of the skin or in the lining of internal organs

CIMP: CpG island methylator phenotype particularly associated with mutant IDH enzyme expression

Clonal evolution: Tumour cells evolve from a single transformed cell which accumulates multiple driver mutations that confer increasingly malignant properties on the cell

Complete response: The disappearance of all signs of cancer in response to treatment

Confounding variable: Variables that could influence the outcome of an experiment and therefore donot allow the experimenter to reasonably eliminate plausable alternative explanations for an observed relationship between independent and dependent variables

Copy number variation: Alterations in the number of copies of specific regions of DNA, which can either be deleted or duplicated

Cytology: Examination of cells by microscopic examination for medical purposes

Cytostatic: Inhibition of cell division and reduced cell growth

Cytotoxic: Describes an agent that causes loss of cell viability

Cytotoxin: An agent that kills dividing cells

DNA adduct: A form of DNA damage caused by covalent attachment of a chemical moiety (e.g. addition of methyl or alky group) to DNA bases due to exposure to chemical carcinogens

Driver gene: A gene that carries a driver mutation or is expressed aberrantly in a way that confers a selective growth advantage through epigentic changes

Driver mutation: Mutations which confer a proliferative advantage and increased fitness to a clone

Durable response rate: An efficacy end point used in clinical trials of cancer therapies. Usually defined as the percentage of patients with complete (no tumour evident) or partial response (reduction in tumour size) for a defined period of time (e.g. six months or one year)

Dyskaryosis: Abnormal changes in epithelial cells characterized by hyperchromatic and irregular nuclei

Dysplasia: A state of abnormal cells which can be altered in size, shape, and organization

Eloquent parts of the brain: Areas of the brain that control sensory processing, motor function, and ability for speech and language

Enantiomer: One of two stereoisomers of a chemical that have optical mirror images of each other. These are designated 'R'/'D' and 'S'/'L'

Epigenetic: Alterations in gene expression as a result of non-genetic changes to the genome

Exome sequencing: Sequencing of all protein-coding regions (exomes) of the genome to determine the nucleotide sequence of these regions. Also known as whole exome sequencing (WES)

Extravasation: The exit of cells from blood vessels into the tissues of an organ

Familial cancer: One mutated allele of a cancer causing gene is inherited through the germline which increases the individuals risk of cancer (also known as hereditary cancers)

Fine needle tissue aspirate: Suction of tissue through a needle to obtain a biopsy specimen

Fractionation of radiotherapy doses: The total dose of radiation is divided into smaller doses or fractions to reduce the side effects on normal tissues

Frame-shift mutation: A mutation arisng from insertion or deletion of a single (or more) base pairs within the coding DNA that leads to the amino acid sequence being read in an alternative translational reading frame

Haploinsufficient: Of the two copies of a gene, one is mutated whilst the second is wild-type retaining its function

Hayflick limit: The number of times a cell can divide before it stops due to short telomeres thus limiting the replicative capacity of a cell

Heterochromatin: Densely packed chromosomal DNA and associated proteins that is genetically inactive

Histone: Basic (alkaline) proteins found attached to DNA in eukaryotic cells: involved in condensing DNA to chromatin

Holliday junctions: Four arms of double-stranded DNA that are key intermediates in homologous recombination dsDNA break repair

Hyperchromasia: Dark staining, when applied to nuclei it usually signifies a high DNA content

Immune related adverse events: Immune effects on normal tissue resulting from misdirected stimulation (e.g. due to drug administration) of the immune system

Immunoediting: Evolution of tumours such that the tumour cells are no longer effectively recognized and killed by the immune system

Immunogenicity: The ability of a molecule or a substance (e.g. antigen, epitope) to elicit an immune response in an organism

Initiating mutation: A mutation that ocurs in a normal cell that permanently alters its genetic material but does not immediatley alter its phenotype

Inter-tumour heterogeneity: Differences between tumours of the same type in different patients

Intra-tumour heterogeneity: Differences between cancer cells within a tumour in the same patient

Intravasation: The entry of cells into blood vessels

Macrometastasis: Large, visible tumours that arise at secondary sites

Mesenchymal glioma: A subtype of glioblastoma characterized by a mutant neurofibromin 1 (*NF1*) gene and high activity in the tumour necrosis factor and *NF-kB* pathways

Metabolome: Full complement of metabolites present in a cell, tissue, or organism

Micrometastasis: Small numbers of cancer cells that migrate to distant locations/secondary sites that remain in a dormant state for a period of time at that site

Microsatellite instability: An increased propensity for mutational alterations in short tandem repeat sequences (microsatellite) caused by defects in DNA mismatch repair mechanisms

Missense mutation: A single nucleotide substitution (point mutation) that changes a codon specifying one amino acid to a different amino acid

Morbidity: A state of sickness or lack of well-being

Mutator phenotype: Refers to the increased rate at which cancer cells acquire mutations. This phenotype results from mutations in genes that maintain genome stability

Necroptosis: A programmed form of necrosis, or inflammatory cell death

Necrosis: Changes due to cell death resulting from irreversible damage involving rupture of cell

Neoantigens: Somatic mutations in the cancer genome that result in the production of mutant proteins that are specific to the tumor and not expressed in normal cells. Can elicit an immune response

Neoplasms: A new growth formed of a mass of cells proliferating abnormally. Neoplasms can be malignant or benig

Non-canonical functions: Not the normal or main functions

Nonsense mutation: A single nucleotide substitution that changes a codon specifying an amino acid to a stop codon, resulting in premature termination of translation

Objective drug response: One of the primary clinical trial end points for drug discovery based on unbiased measurable changes

Objective response rate: An efficacy end point used in clinical trials of cancer therapies in solid tumours. Usually defined as the percentage of patients who experience at least 30% decrease in tumour diameter on an imaging scan

Oncogene addiction: The dependency of a tumour on a single oncogene for its growth and survival

OncomiR: Cancer associated microRNA with oncogenic properties

Palpable: Capable of being felt

Parenchyma: The functional tissue of the organ, distinguished from the stroma which refers to the structural tissue of the organ

Passenger mutation: Mutations which do not confer proliferative advantage but are acquired by the genome and carried along during clonal expansion. Their significance in cancer is less well-defined

Penetrance: The frequency with which a a specific genotype manifests the corresponding phenotype. For example, in cases of complete penetrance, a 100% of the individuals in a population who carry a specific genotype express the related phenotype

Permissive (temperature): Temperature at which the encoded protein product folds and therefore functions properly. At the non-permissive temperature, protein does not fold properly and therefore protein function is lost

Plasticity (cell): The ability of cells to switch from one phenotype to a different phenotype

Pleomorphism: Variability in size, morphology, and appearance

Polymorphism: Two or more variants of a sequence (e.g. alleles, sequence variants) that occur at significant frequencies in a population, typically at a frequency of more than 1% in a population

Premetastatic niche: An environment in a potential secondary site that is conducive to supporting metastatic cell survival

Promoting mutation: Mutations that occur in cells that have already been initiated, thus further mutations promote proliferation and cell survival

Proneural: A subtype of glioblastoma characterized by a mutant IDH and a CpG island methylated phenotype and associated with a more favourable prognosis

Recall bias: Participants in a retrospective study either forget to report particular information or under or over report information leading to errors in the data collated

Recurrent mutation: Driver mutations are recurrent, that is, the same gene is affected in multiple individuals within a cohort. In contrast, passenger mutations are random and would not be expected to affect the same gene in different people

Relative risk: A measure of the risk of a certain event happening in one group (e.g. cancer) compared to the risk of the same event happening in another group. A relative risk (RR) of 1 means no difference between the two groups being compared, RR higher than 1 signifies increased risk and RR lower than 1 signifies reduced risk

Retrovirus: RNA viruses which insert a DNA copy of their genome into the host cell in order to replicate

SASP: Senescence-associated secretory phenotype whereby senescent cells secrete an array of cytokines, growth factors and proteases

Selection bias: A form of error (bias) that results from selection of an unrepresentative sample and can lead to overestimating or underestimating the outcome being measured

Self-renewal: Property of stem cells; ability to generate descendants that retain stemness characteristics

Senescence: Arrest of cell division in response to cellular stress

Significantly mutated genes: Genes that display higher mutation frequencies (than background mutation rates) when multiple tumours are analysed for gene mutations (e.g. indels, single nucleotide variation, etc.) using statistical and computational methods. Analysis separates out genes involved in driving disease (driver mutations) from the passenger mutations

Sporadic cancer: Cancers that arise as a consequence of mutations in somatic cells and are not inherited through the germline

Stem cell niche: An area of tissue that houses stem cells, providing them with a specific microenvironment that supports their undifferentiated and self-renewable state

Stereotactic: A computer-assisted 3D imaging system of a tumour to make surgical removal or radiotherapy more accurate

Stroma: The tissue forming the supportive matrix rather than the functional part of an organ

Subtelomeric region: Region of a chromosome immediately centromeric to the telomere repeats

Symmetrical cell division: Method of cell division in which two identical daughter cells are produced

Telomere crisis: A state of critically short telomeres usually resulting in cell death or activation of a telomre maintenance mechanism

Therapeutic index: A comparison of efficacy versus toxicity of a drug. A high therapeutic index is preferable to a low one

Transitions: A DNA base change in which a purine is substituted for another purine or a pyrimidine base substituted for another pyrimidine

Transversion: A DNA base change in which a purine is substituted for a pyrimidine or a pyrimidine substituted for a purine

Tropism: The specificity of a virus for a particular host tissue

Tumour: A swelling formed of a mass of cells resulting from abnormal proliferation

Tumour associated antigens: Antigens expressed by tumour cells and by normal cells, however, expression of the antigen is higher in cancer cells than in normal cells. Can elicit an immune response

Tumour microenvironment: The cellular and chemical environment in which the tumour cells exist including blood vessels, fibroblasts, immune cells, and extracellular matrix

Tumourigenic: Tumour-generating; in the context of the cancer stem cell, able to generate a new tumour that comprises of tumourigenic and non-tumourigenic cell populations. This property is also called tumour initiating capability or tumour propagating capability

Vascular mimicry: The ability of tumour cells to acquire endothelial like properties and form microvascular channels

Whole genome sequencing: Method for determining the nucleotide sequence of the full genome (protein-coding and non-coding regions)

REFERENCES

Chapter 1

Bailey, M. H., et al. (2018). 'Comprehensive characterization of cancer driver genes and mutations'. Cell, 173(2): 371–85.e318.

Berenblum, I. and Shubik, P. (1947). 'The role of croton oil applications, associated with a single painting of D and a carcinogen, in tumour induction of the mouse's skin'. British Journal of Cancer, 1(4): 379–82.

Bonnet, D and Dick, J. E. (1997). 'Human acute myeloid leukemia is organized as a hierarchy that originates from a primitive hematopoietic cell'. Nature Medicine, 3(7): 730–7.

Stephens, P. J., et al. (2012). 'The landscape of cancer genes and mutational processes in breast cancer'. Nature, 486: 400–4.

Vogelstein, B., et al. (2013). 'Cancer genome landscapes'. Science, 339(6127): 1546–58.

Vogelstein, B. and Kinzler, K. W. (1993). 'The multistep nature of cancer'. Trends in Genetics, 9(4): 138–41.

Vogt, P. K. (2012). 'Retroviral oncogenes: a historical primer'. Nature Reviews Cancer, 12(9): 639–48.

Weinberg, R. A. (2014). The Biology of Cancer. Second Edition. Garland Science. Abingdon, UK.

Chapter 2

Coudray, N., Ocampo, P. S., Sakellaropoulos, T., Narula, N., Snuderl, M., Fenyö, D., Moreira, A. L., Razavian, N., and Tsirigos, A. (2018). 'Classification and mutation prediction from non-small cell lung cancer histopathology images using deep learning'. Nature Medicine, 24(10): 1559–67.

Fouad, Y. A. and Aaner, C. (2017). 'Revisiting the hallmarks of cancer'. American Journal of Research, 7(5): 1016–36.

Hanahan, D. (2014). 'Cancer wars 2. Rethinking the war on cancer'. Lancet, 383: 558–63.

Orchard, G. E. and Nation, B. R. (eds) (2011). Histopathology. Oxford University Press. ISBN 9780198717331.

Winkler, Frank (2017). 'Hostile takeover: how tumours hijack pre-existing vascular environments to thrive'. The Journal of Pathology, 242: 267–72.

Zakari, N., Yusoff, N. M., Zakaria, Z., Lim, M. N., Baharuddin, P. J, Fakiruddin, K. S., and Yahaya, B. (2015). 'Human non-small cell lung cancer expresses putative cancer stem cell markers and exhibits the transcriptomic profile of multipotent cells'. BMC Cancer, 15: 84–100.

Chapter 3

Ahmad, A. S., et al. (2015). 'Trends in the lifetime risk of developing cancer in Great Britain: comparison of risk for those born from 1930 to 1960'. British Journal of Cancer, 112(5): 943–7.

Alexandrov, L. B., et al. (2013). 'Signatures of mutational processes in human cancer'. Nature, 500(7463): 415–21.

Alexandrov, L. B., et al. (2016). 'Mutational signatures associated with tobacco smoking in human cancer'. Science, 354(6312): 618–22.

Amos, C. I., et al. (2008). 'Genome-wide association scan of tag SNPs identifies a susceptibility locus for lung cancer at 15q25.1'. Nature Genetics, 40: 616.

Bray, F., et al. (2018). 'Global cancer statistics 2018: GLOBOCAN estimates of incidence and mortality worldwide for 36 cancers in 185 countries'. CA: A Cancer Journal for Clinicians, 68(6): 394–424.

Carbone, M., et al. (2018). 'Consensus report of the 8 and 9th Weinman Symposia on gene x environment interaction in carcinogenesis: Novel opportunities for precision medicine'. Cell Death & Differentiation, 25(11): 1885–904.

Chandler, M. R., et al. (2016). 'A review of whole-exome sequencing efforts toward hereditary breast cancer susceptibility gene discovery'. Human Mutation, 37(9): 835–46.

Demeyer, D., et al. (2016). 'Mechanisms linking colorectal cancer to the consumption of (processed) red meat: a review'. Critical Reviews in Food Science & Nutrition, 56(16): 2747–66.

Fanale, D., et al. (2012). 'Breast cancer genome-wide association studies: there is strength in numbers'. Oncogene, 31(17): 2121–8.

Flemer, B., et al. (2017). 'Tumour-associated and non-tumour-associated microbiota in colorectal cancer'. Gut, 66(4): 633–43.

Hall, J., et al. (1990). 'Linkage of early-onset familial breast cancer to chromosome 17q21'. Science, 250(4988): 1684–9.

Hayward, N. K., et al. (2017). 'Whole-genome landscapes of major melanoma subtypes'. Nature, 545(7653): 175–80.

Hecht, S. S. (2012). 'Lung carcinogenesis by tobacco smoke'. International Journal of Cancer, 131(12): 2724–32.

Helleday, T., et al. (2014). 'Mechanisms underlying mutational signatures in human cancers'. Nature Reviews Genetics, 15: 585–98.

Hung, R. J., et al. (2008). 'A susceptibility locus for lung cancer maps to nicotinic acetylcholine receptor subunit genes on 15q25'. Nature, 452(7187): 633–7.

International Agency for Research on Cancer (IACR) (2014). 'World Cancer Report 2014'. World Health Organization, Geneva.

Konstantinov, S. R. (2017). 'Diet, microbiome, and colorectal cancer'. Best Practice & Research Clinical Gastroenterology, 31(6): 675–81.

Lamprecht, S. A. and Lipkin, M. (2003). 'Chemoprevention of colon cancer by calcium, vitamin D and folate: molecular mechanisms'. Nature Reviews Cancer, 3: 601–14.

McKay, J. D., et al. (2017). 'Large-scale association analysis identifies new lung cancer susceptibility loci and heterogeneity in genetic susceptibility across histological subtypes'. Nature Genetics, 49(7): 1126–32.

O'Keefe, S. J. D. (2016). 'Diet, microorganisms and their metabolites, and colon cancer'. Nature Reviews Gastroenterology & Hepatology, 13: 691–706.

Petljak, M. and Alexandrov, L. B. (2016). 'Understanding mutagenesis through delineation of mutational signatures in human cancer'. Carcinogenesis, 37(6): 531–40.

Renehan, A. G., et al. (2015). 'Adiposity and cancer risk: new mechanistic insights from epidemiology'. Nature Reviews Cancer, 15: 484–98.

Royston, K.J., Adedokun, B., and Olopade, O.I. (2019). 'Race, the microbiome and colorectal cancer'. World Journal of Gastrointestinal Oncology, 11(10): 773–87.

Scoccianti, C. E. A. (2013). 'Recent evidence on alcohol and cancer epidemiology'. Future Oncology, 9(9): 1315–22.

Shiovitz, S. and Korde, L. A. (2015). 'Genetics of breast cancer: a topic in evolution'. Annals of Oncology, 26(7): 1291–9.

Simonds, N. I., et al. (2016). 'Review of the gene-environment interaction literature in cancer: what do we know?' Genetic Epidemiology, 40(5): 356–65.

Smittenaar, C. R., et al. (2016). 'Cancer incidence and mortality projections in the UK until 2035'. British Journal of Cancer, 115(9): 1147–55.

Teugels, E. and De Brakeleer, S.(2017). 'An alternative model for (breast) cancer predisposition'. npj Breast Cancer, 3(1): 13.

The Chek Breast Cancer Consortium (2002). 'Low-penetrance susceptibility to breast cancer due to CHEK2*1100delC in noncarriers of BRCA1 or BRCA2 mutations'. Nature Genetics, 31: 55–9.

Thorgeirsson, T. E., et al. (2008). 'A variant associated with nicotine dependence, lung cancer and peripheral arterial disease'. Nature, 452: 638–42.

Wen, L., et al. (2016). 'Contribution of variants in CHRNA5/A3/B4 gene cluster on chromosome 15 to tobacco smoking: from genetic association to mechanism'. Molecular Neurobiology, 53(1): 472–84.

Wu, S., et al. (2016). 'Substantial contribution of extrinsic risk factors to cancer development'. Nature, 529(7584): 43–7.

World Cancer Research Fund International/American Institute for Cancer Research (2018). 'Continuous Update Project: Diet, Nutrition, Physical Activity and Cancer: a Global Perspective'. Available at https://www.wcrf.org/dietandcancer.

Chapter 4

Allard, B., et al. (2018). 'Immuno-oncology-101: overview of major concepts and translational perspectives'. Seminars in Cancer Biology, 52: 1–11.

Chen, D. S. and Mellman, I. (2013). 'Oncology meets immunology: the cancer-immunity cycle'. Immunity, 39(1): 1–10.

Chen, Y., Peng, Y., Fan, S. et al. (2017). A double dealing tale of p63: an oncogene or a tumor suppressor. Cellular & Molecular Life Sciences, 75: 965–73.

Engeland, K. (2017). 'Cell cycle arrest through indirect transcriptional repression by p53: I have a DREAM'. Cell Death & Differentiation, 25: 114–32.

Goswami, K. K., et al. (2017). 'Tumor promoting role of anti-tumor macrophages in tumor microenvironment'. Cellular Immunology, 316: 1–10.

Horikawa, I., Park, K-Y., Isogaya, K., Hiyoshi, Y., Li, H., Anami, K., Robles, A. I., Mondal, A. M., Fujita, K., Serrano, M., and Harris, C. C. (2017). 'Δ133p53 represses p53-inducible senescence genes and enhances the generation of human induced pluripotent stem cells'. Cell Death and Differentiation, 24: 1017–28.

Koivunen, P., Sungwoo, L., Duncan, C. G., Lopez, G., Lu, G., Ramkissoon, S., et al. (2012). 'Transformation by the (R)-enantiomer of 2-hydroxyglutarate linked to EGLN activation'. Nature, 483: 484–8.

Mojic, M., et al. (2018). 'The dark side of IFN-γ: its role in promoting cancer immunoevasion'. International Journal of Molecular Sciences, 19(1): 89. doi:10.3390/ijms19010089

Moulder, D. E., Hatoum, D., Tay, E., Lin, Y., and McGowan, E. M. (2018). 'The roles of p53 in mitochondrial dynamics and cancer metabolism: the pendulum between survival and death in breast cancer'. Cancers, 10(189): 1–22.

Nagarsheth, N., et al. (2017). 'Chemokines in the cancer microenvironment and their relevance in cancer immunotherapy'. Nature Reviews Immunology, 17: 559–72.

Olive, K. P., Tuveson, D. A., Ruhe, Z. C., Yin, B., Willis, N. A., Bronson, R. T., Crowley, D., and Jacks, T. (2004). 'Mutant p53 Gain of Function in two mouse models of Li-Fraumeni Syndrome'. Cell, 119: 847–60.

O'Neil, N. J., Bailey, M. L. and Hieter, P. (2017). 'Synthetic lethality and cancer'. Nature Reviews Genetics, 18(10): 613–23.

Pandya, P. H., Murray, M. E., Pollok, K. E., & Renbarger, J. L. (2016). 'The immune system in cancer pathogenesis: Potential therapeutic approaches'. Journal of immunology research, 2016:4273943. doi:10.1155/2016/4273943.

Rodríguez, N., Peláez, A., Barderas, R., and Domínguez, G. (2018). 'Clinical implications of the deregulated TP73 isoforms expression in cancer'. Clinical & Translational Oncology, 20: 827–36.

Singh, S. S., et al. (2018). 'Dual role of autophagy in hallmarks of cancer'. Oncogene, 37(9): 1142–58.

Wang, M., et al. (2017). 'Role of tumor microenvironment in tumorigenesis'. Journal of Cancer, 8(5): 761–73.

Chapter 5

Blandino, G. and Di Agostino, S. (2018). 'New therapeutic strategies to treat human cancers expressing mutant p53 proteins'. Journal of Experimental & Clinical Cancer Research, 37: 30–43.

Brown, J. S. and Banerji, U. (2017). 'Maximising the potential of AKT inhibitors as anti-cancer treatments'. Pharmacology & Therapeutics, 172: 101–15.

Brudno, J. N. and Kochenderfer, J. N. (2017). 'Chimeric antigen receptor T-cell therapies for lymphoma'. Nature Reviews Clinical Oncology, 15: 31–46.

Bykov, V., Eriksson, S. E., Bianchi, J., and Wiman, K. G. (2018). 'Targeting mutant p53 for efficient cancer therapy'. Nature Reviews Cancer, 18(2): 89–102.

Clarke, J. M. and Hurwitz, H. I. (2013). 'Understanding and targeting resistance to anti-angiogenic therapies'. Journal of Gastrointestinal Oncology, 4(3): 253–63.

Diesendruck, Y. and Benhar, I. (2017). 'Novel immune check point inhibiting antibodies in cancer therapy—Opportunities and challenges'. Drug Resistance Updates, 30: 39–47.

Farkona, S., et al. (2016). 'Cancer immunotherapy: the beginning of the end of cancer?' BMC Medicine, 14(1): 73. doi:10.1186/s12916-016-0623-5

Finn, O. J. (2017). 'The dawn of vaccines for cancer prevention'. Nature Reviews Immunology, 18: 183–94.

June, C. H., et al. (2018). 'CAR T cell immunotherapy for human cancer'. Science, 359(6382): 1361–5.

Khalil, D. N., et al. (2016). 'The future of cancer treatment: immunomodulation, CARs and combination immunotherapy'. Nature Reviews Clinical Oncology, 13(5): 273–90.

Losman, J-A. and Kaelin, W. G. (2013). 'What a difference a hydroxyl makes: mutant IDH, (R)-2-hydroxyglutarate and cancer'. Genes & Development, 27(8): 836–52.

Miyamoto, Y., Suyama, K., and Baba, H. (2017). 'Recent advances in targeting the EGFR signaling pathway for the treatment of metastatic colorectal cancer'. International Journal of Molecular Sciences, 18: 752–67.

O'Hare, T., Deininger, M. W. M., Eide, C. A., Clackson, T., and Druke, B. J. (2010). 'Targeting the BCR-ABL signaling pathway in therapy-resistant Philadelphia chromosome-positive leukemia'. Clinical Cancer Research, 7(2): 212–21.

O'Donnell, J.S., Teng, W.L., and Smyth, M.J. (2019). 'Cancer immunoediting and resistance to T cell-based immunotherapy'. Nature Reviews Clinical Oncology, 16(3): 151–67.

Ribas, A. and Wolchok, J. D. (2018). 'Cancer immunotherapy using checkpoint blockade'. Science, 359(6382): 1350–5.

Rimkus, T. K. Carpenter, R. L., Qasem, S., Chan, M., and Lo, H-W. (2016). 'Targeting the sonic hedgehog signaling pathway: review of smoothened and GLI inhibitors'. Cancers, 8: 22–45.

Waitkus, M. S., Diplas, B. H., and Yan, H. (2018). 'Biological role and therapeutic potential of IDH mutations in cancer'. Cancer Cell, 34: 1–10.

Yuan, Y., Liu, L., Chen, H., Wang, Y., Xu, Y., Mao, H., Li, J., Mills, G. B., Shu, Y., Li, L., and Liang, H. (2016). 'Comprehensive characterization of molecular differences in cancer between male and female patients'. Cancer Cell, 29(5): 711–22.

Chapter 6

Banchereau, J. and Palucka, K. (2017). 'Cancer vaccines on the move'. Nature Reviews Clinical Oncology, 15: 9–10.

Bilsland, A. E., et al. (2016). 'Virotherapy: cancer gene therapy at last?' F1000Research 5: F1000 Faculty Rev-2105.

Borrebaeck, C. A. K. (2017). 'Precision diagnostics: moving towards protein biomarker signatures of clinical utility in cancer'. Nature Reviews Cancer, 17: 199–204.

Brock, A. and Huang,S. (2017). 'Precision oncology: between vaguely right and precisely wrong'. Cancer Research, 77(23): 6473–9.

Byers, L. A., et al. (2013). 'An epithelial–mesenchymal transition gene signature predicts resistance to EGFR and PI3K inhibitors and identifies Axl as a therapeutic target for overcoming EGFR inhibitor resistance'. Clinical Cancer Research, 19(1): 279–90.

Cao, M.-X., et al. (2017). 'The crosstalk between lncRNA and microRNA in cancer metastasis: orchestrating the epithelial-mesenchymal plasticity'. Oncotarget, 8(7): 12472–83.

Chaffer, C. L., et al. (2016). 'EMT, cell plasticity and metastasis'. Cancer & Metastasis Reviews, 35(4): 645–54.

Collins, D. C., et al. (2017). 'Towards precision medicine in the clinic: from biomarker discovery to novel therapeutics'. Trends in Pharmacological Sciences, 38(1): 25–40.

Conry, R. M., et al. (2018). 'Talimogene laherparepvec: First in class oncolytic virotherapy'. Human Vaccines & Immunotherapeutics, 14(4): 839–46.

Cornu, T. I., et al. (2017). 'Refining strategies to translate genome editing to the clinic'. Nature Medicine, 23: 415–23.

Cyranoski, D. (2016). 'CRISPR gene-editing tested in a person for the first time'. Nature, 539(479). doi: 10.1038/nature.2016.20988.

Dagogo-Jack, I. and Shaw, A. T. (2017). 'Tumour heterogeneity and resistance to cancer therapies'. Nature Reviews Clinical Oncology, 15: 81–94.

Dongre, A. and R. A. Weinberg (2019). 'New insights into the mechanisms of epithelial–mesenchymal transition and implications for cancer'. Nature Reviews Molecular Cell Biology, 20(2): 69–84.

Draghi, A., et al. (2018). 'Acquired resistance to cancer immunotherapy'. Seminars in Immunopathology. doi: 10.1007/s00281-018-0692-y.

Dummer, R., et al. (2017). 'Combining talimogene laherparepvec with immunotherapies in melanoma and other solid tumors'. Cancer Immunology, Immunotherapy, 66(6): 683–95.

Hu, Z., et al. (2017). 'Towards personalized, tumour-specific, therapeutic vaccines for cancer'. Nature Reviews Immunology, 18: 168–82.

Janiszewska, M., et al. (2015). 'In situ single-cell analysis identifies heterogeneity for PIK3CA mutation and HER2 amplification in HER2-positive breast cancer'. Nature Genetics, 47: 1212–19.

Keller, L., and Pantel, K. (2019). 'Unravelling tumour heterogeneity by single-cell profiling of circulating tumour cells'. Nature Reviews, 19: 553–67.

Konieczkowski, D. J., et al. (2018). 'A convergence-based framework for cancer drug resistance'. Cancer Cell, 33(5): 801–15.

Lambert, A. W., et al. (2017). 'Emerging biological principles of metastasis'. Cell, 168(4): 670–91.

Lawler, S. E., et al. (2017). 'Oncolytic viruses in cancer treatment: A review'. JAMA Oncology, 3(6): 841–9.

Lebrun, J.-J. (2012). 'The dual role of TGFβ in human cancer: from tumor suppression to cancer metastasis'. Molecular Biology, 52(1): 39–55.

Li, H., et al. (2018). 'A serum microRNA signature predicts trastuzumab benefit in HER2-positive metastatic breast cancer patients'. Nature Communications, 9(1): 1614. doi:10.1038/s41467-018-03537-w

Lobb, R. J., et al. (2017). 'Exosomes: Key mediators of metastasis and pre-metastatic niche formation'. Seminars in Cell & Developmental Biology, 67: 3–10.

Marquardt, S., et al. (2018). 'Emerging functional markers for cancer stem cell-based therapies: Understanding signaling networks for targeting metastasis'. Seminars in Cancer Biology. doi: 10.1016/j.semcancer.2018.06.006.

Nair, M., et al. (2018). 'Cancer molecular markers: A guide to cancer detection and management'. Seminars in Cancer Biology, 52(1): 39–55.

Nikolaou, M., et al. (2018). 'The challenge of drug resistance in cancer treatment: a current overview'. Clinical & Experimental Metastasis, 35(4): 309–18.

Nicolini, A., et al. (2017). 'Prognostic and predictive biomarkers in breast cancer: Past, present and future'. Seminars in Cancer Biology, 52(1): 56–73.

Ott, P. A., et al. (2017). 'An immunogenic personal neoantigen vaccine for patients with melanoma'. Nature, 547(7662): 217–21.

Peinado, H., et al. (2017). 'Pre-metastatic niches: organ-specific homes for metastases'. Nature Reviews Cancer, 17: 302.

Pietilä, M., et al. (2016). 'Whom to blame for metastasis, the epithelial–mesenchymal transition or the tumor microenvironment?' Cancer Letters, 380(1): 359–68.

Roth, T. L., et al. (2018). 'Reprogramming human T cell function and specificity with non-viral genome targeting'. Nature, 559(7714): 405–9.

Sahin, U., et al. (2017). 'Personalized RNA mutanome vaccines mobilize poly-specific therapeutic immunity against cancer'. Nature, 547: 222–6.

Sahin, U. and Türeci, Ö. (2018). 'Personalized vaccines for cancer immunotherapy'. Science, 359(6382): 1355–60.

Shibue, T. and Weinberg, R. A. (2017). 'EMT, CSCs, and drug resistance: the mechanistic link and clinical implications'. Nature Reviews Clinical Oncology, 14: 611–29.

Siravegna, G., et al. (2015). 'Clonal evolution and resistance to EGFR blockade in the blood of colorectal cancer patients'. Nature Medicine, 21: 795–801.

Song, W., et al. (2017). 'Translational Significance for Tumor Metastasis of Tumor-Associated Macrophages and Epithelial–Mesenchymal Transition'. Frontiers in Immunology, 8: 1106.

Steeg, P. S. (2016). 'Targeting metastasis'. Nature Reviews Cancer, 16(4): 201–18.

Suarez-Carmona, M., et al. (2017). 'EMT and inflammation: inseparable actors of cancer progression'. Molecular Oncology, 11(7): 805–23.

Tellez-Gabriel, M., et al. (2016). 'Tumour heterogeneity: the key advantages of single-cell analysis'. International Journal of Molecular Sciences, 17(12): 2142. doi:10.3390/ijms17122142.

Tsoucas, D. and Yuan, G.-C. (2017). 'Recent progress in single-cell cancer genomics'. Current Opinion in Genetics & Development, 42: 22–32.

van Staalduinen, J., et al. (2018). 'Epithelial–mesenchymal-transition-inducing transcription factors: new targets for tackling chemoresistance in cancer?'. Oncogene, 37(48): 6195–211.

Vasan, N., Baselga, J., and Hyman, D.M. (2019). 'A view on drug resistance and cancer'. Nature, 575(7782): 299–309.

Williams, E.D., Gao, D, Redfern, A., and Thompson, E.W. (2019). 'Controversies around epithelial-mesenchymal plasticity in cancer metastasis'. Nature, 19(12): 716–32.

Wu, G., et al. (2017). 'Overcoming treatment resistance in cancer: current understanding and tactics'. Cancer Letters, 387: 69–76.

INDEX